MODERN ISRAELI   Israeli Air Force

Thoma   tos)

MODERN ISRAELI AIR POWER

Aircraft and Units of the Israeli Air Force

Thomas Newdick (text) and **Ofer Zidon** (photos)

HARPIA
PUBLISHING +

Copyright © 2013 Harpia Publishing L.L.C. & Moran Publishing L.L.C. Joint Venture
2803 Sackett Street, Houston, TX 77098-1125, U.S.A.
miap@harpia-publishing.com

All rights reserved.

No part of this publication may be copied, reproduced, stored electronically or transmitted in any manner or in any form whatsoever without the written permission of the publisher.

Consulting and inspiration by Kerstin Berger
Drawings by Tom Cooper
Map by James Lawrence
Layout by Norbert Novak

Harpia Publishing, L.L.C. is a member of

Printed at Grasl Druck & Neue Medien, Austria

ISBN 978-0-9854554-2-2

חיל האוויר והחלל הישראלי

Contents

Introduction 7

Acknowledgements 8

Abbreviations 9

Chapter 1: A Brief History 11

Chapter 2: IASF Combat Operations in the 21st Century 19

Chapter 3: International Activities in the 21st Century 33

Chapter 4: IASF Aircraft Types 39

Chapter 5: IASF Ordnance (co-written by Sariel Stiller) 69

Chapter 6: Hatzerim Air Base 85

Chapter 7: Hatzor Air Base 103

Chapter 8: Nevatim Air Base 113

Chapter 9: Ovda Air Base 129

Chapter 10: Palmachim Air Base 135

Chapter 11: Ramat David Air Base 147

Chapter 12: Ramon Air Base 165

Chapter 13: Sde Dov Air Base 181

Chapter 14: Tel Nof Air Base 189

Chapter 15: Air Defense Command 209

Appendix I: IASF Order of Battle, 2013 215

Appendix II: IASF Colours and Markings 221

Appendix III: Current Hebrew Type Names 247

Bibliography 249

Index 251

חיל האוויר והחלל הישראלי

Introduction

In recent years the Israeli Air Force (officially the Israel Air and Space Force – IASF) has undergone significant changes in terms of both structure and hardware, gaining combat experience while participating in the 2006 Ebb and Flow and 2008–09 Cast Lead operations in Gaza and in the July 2006 conflict in Lebanon, and improving its organisation in order to further shorten reaction times and increase the tempo of operations. During this timeframe new aircraft types have entered service, while some older types have been withdrawn. New squadrons and bases have been activated while others have been closed. The unmanned fleet has taken huge strides, the F-16I fleet has expanded to become the backbone of the IASF's fighter force and the transport wing has moved from Lod to its new location at Nevatim in the southern Negev Desert.

This book includes coverage of each of the IASF's different branches: fighter, transport, special operations, rotary-wing attack and assault, the unmanned fleet and the Air Force Academy. The current status of each element within the IASF is discussed in detail in the context of the changes that have taken place over the past decade. The result is the most comprehensive account of the modern Israeli Air Force, its organisation, weapons and capabilities.

Thomas Newdick
August 2013

Acknowledgements

For their invaluable cooperation in the preparation of this book, the author and Ofer Zidon wish to extend their thanks to Adi Chazak, Asher Roth, Ella Shechter, Erez Sirotkin, Sariel Stiller, Ilan Warshai and Ra'anan Weiss.

In addition, Ofer Zidon wishes to thank those who helped him with photos and with photography sessions at Israeli Air Force bases and installations: Major Ofer B., Major Roy B., Nechemia Gershuni, Carmel Horowitz, Yuval K., P. Panagiotopoulos and Yissachar Ruas.

For the provision of access to the Israeli Air Force and its units, the author and Ofer Zidon are indebted to the Israeli Air Force communication and media desk and the Israel Defense Forces Spokesperson.

Last but not the least, this book would not have been possible without the assistance of the squadrons of the Israeli Air Force and the personnel at its various bases.

חיל האוויר והחלל הישראלי

Abbreviations

AA	anti-aircraft
AAM	air-to-air missile
ABM	anti-ballistic missile
AEW	airborne early warning
AFV	armoured fighting vehicle
AMRAAM	Advanced Medium-Range Air-to-Air Missile
AOTU	Advanced Operational Training Unit
ATC	Advanced Training Center
CAEW	Conformal Airborne Early Warning
CFT	conformal fuel tank
CSAR	combat search and rescue
EAF	Egyptian Air Force (see also UARAF)
ECM	electronic countermeasures
ELINT	electronic intelligence
EO	electro-optical
FLIR	forward-looking infrared
GP	general-purpose (bomb)
GPS	Global Positioning System
HAS	hardened aircraft shelter
IAI	Israel Aircraft Industries and, from November 2006, Israel Aerospace Industries
IASF	Israel Air and Space Force
IDF	Israel Defense Forces
IDF/AF	Israel Defense Forces/Air Force
IMI	Israel Military Industries
IR	infra-red
JDAM	Joint Direct Attack Munition
QRA	quick reaction alert
LANTIRN	Low-Altitude Navigation and Targeting Infrared for Night
LGB	laser-guided bomb
LOROP	long-range oblique photography
OTU	Operational Training Unit
MIA	missing in action
PLO	Palestine Liberation Organization
PoW	prisoner of war
RAF	Royal Air Force (of the United Kingdom)
RJAF	Royal Jordanian Air Force
SAM	surface-to-air missile
SATCOM	satellite communication(s)
SEAD	suppression of enemy air defences
SIGINT	signals intelligence
SyAAF	Syrian Arab Air Force
UAV	unmanned aerial vehicle
UARAF	United Arab Republic Air Force, official name of the Egyptian Air Force between 1958 and 1972
UN	United Nations
UNIFIL	United Nations Interim Force in Lebanon
USAAF	United States Army Air Force
VISINT	visual intelligence

Chapter 1

A BRIEF HISTORY

Foundation and the 1948 War of Independence

The Israeli Air Force was originally established as an independent arm of the Israel Defense Forces, and was correspondingly frequently known as the Israel Defense Forces/Air Force (IDF/AF) until 2005, when it was reorganised under a new title as an independent branch of the military – the Israel Air and Space Force (IASF).

The IDF/AF came into being primarily on the basis of foreign and some Jewish personnel that had served with the Royal Air Force (RAF) and US Army Air Force (USAAF) during World War II, and was organised around infrastructure and airfields constructed by the British during their mandate in Palestine between 1922 and 1948. At first, it deployed a miscellany of civilian and military aircraft obtained from a variety of different sources. Unsurprisingly, during its first few years of operations, the IDF/AF was strongly influenced by the RAF's organisational and operational methods, and by its esprit de corps. In turn, it was precisely due to this influence that it soon became regarded as the most professional branch of the IDF.

The early thinking behind the purchase of IDF/AF hardware was simple: to obtain whatever was available from any supplier that was willing to sell. Using this policy, the IDF/AF brought together a motley collection of ex-British Auster Autocrats, Czech-built Avia S-199s, ex-USAAF Boeing B-17G Flying Fortresses, ex-Czechoslovak Supermarine Spitfires, ex-Swedish and Italian North American P-51 Mustangs, and ex-French de Havilland Mosquitoes, among many other aircraft types.

Experiences from both the 1948 War of Independence and from World War II served to crystallise the IDF/AF's primary aim: establishment of air superiority over the battlefield and provision of support for its own ground and naval forces. Because of the country's small size and the vulnerability of its population centres, the idea was born to project any war outside of Israel's borders. According to official Israeli histories, one of the first fighter actions of the newly established IDF/AF included an interception of Egyptian Douglas C-47s that had been converted as bombers, using an S-199 Sakeen (knife), in the course of which the IDF/AF posted claims for two Egyptian aircraft shot down.

Bitter air battles continued despite several UN-sponsored ceasefires, and culminated in an Israeli offensive during late 1948 and early 1949, in the course of which the IDF/AF established air superiority within Israeli airspace.

The IDF/AF's secondary role – ground attack and close air support – was also developed during the course of the 1948 war. Using light civil aircraft types such as the Auster, Noorduyn Norseman, Beech Bonanza, Avro Anson and other types, the IDF/AF accomplished bombing missions against Arab and Palestinian forces by the simple expedient of throwing 50kg (110lb) bombs and hand grenades through the aircraft's windows and doors. The same aircraft were also used for supply drops of food and ammunition to remote towns and for the escort of supply convoys.

The IDF/AF's third role – strategic strike – was also developed during the 1948 conflict. A good example of this concept was the use of three B-17Gs to bomb Cairo, the Egyptian capital, during their ferry flight to Israel.

1949–56 and the Sinai Campaign

The years between the 1948 War of Independence and the 1956 Suez campaign were used to strengthen the foundations of a modern air force. During that period a new organisational structure was introduced, the existing British infrastructure was enhanced and new aircraft types were fielded. The IDF/AF's first jet fighter, the British-made Gloster Meteor arrived in Israel in June 1953. 1955 saw the beginning of the 'French era' in the IDF/AF. In October 1955 the IDF/AF purchased 24 Dassault Ouragans, 12 Dassault Mystère IVAs and three Nord Noratlas transports. In the following year another 24 Mystère IVAs were purchased and a new airspace control system was delivered from France. In late 1956, as part of the preparations for the Suez campaign, another batch of 36 Mystère IVAs and six Ouragans were delivered.

The second Arab-Israeli conflict took place in late October and early November 1956. From the Israeli standpoint, this was a 'pre-emptive' war, launched with the aim of destroying newly delivered Egyptian equipment of Soviet origin – which by now included cutting-edge hardware, such as Mikoyan i Gurevich MiG-17 jet fighters and Ilyushin Il-28 jet bombers – before Egypt became too powerful. Furthermore, it resulted in Israeli occupation of Sinai. As cover, Israel participated in an Anglo-French effort (Operation Musketeer) to regain the newly nationalised Suez Canal and remove the government of Egypt's President Gamal Abdel Nasser. British and French aerial strikes forced the Egyptian military to withdraw from the Sinai Peninsula, leaving it free for Israeli occupation, which ended only after concerted US pressure, in March 1957.

1957–67 and the Six-Day War

The next decade saw the flourishing of the French era, in the course of which the IDF/AF was equipped with almost every new aircraft type that French industry had to offer. IDF/AF bases therefore soon included units flying the Dassault Super Mystère B2 and Dassault Mirage IIICJ – known in Israel as Sambad and Shahak (sky blazer), respectively – light bombers like the Sud Aviation Vautour, transports such as the Noratlas, and Fouga Magister trainers. The rotary-wing fleet was subject to slow but significant expansion through acquisitions of the Sikorsky S-55 and S-58, Sud Aviation Alouette II and Super Frelon, purchased from the US, West Germany and France.

Disputes over the crucial issues of irrigation and water-exploitation projects launched by Israel, Jordan, Syria and Lebanon in the early 1950s culminated in a series of intense clashes along the River Jordan, starting in 1964, and which indirectly provoked the June 1967 War, better known in the West as the Six-Day War. The war broke out on 5 June 1967, after the Israeli military leadership interpreted the build-up of Egyptian military forces in the Sinai Peninsula and the blocking of the Tiran Strait and Gulf of Aqaba to Israeli shipping as a signal that Cairo was preparing to launch an attack on Israel. Between 07.00 and 08.00 Israeli time almost the entire IDF/AF was involved in a carefully planned and rehearsed, concentrated effort to destroy the Egyptian Air Force (EAF, though between 1958 and 1972 officially named as the United Arab Republic Air Force, UARAF). Three hours of sustained Israeli assaults on Egyptian air bases resulted in the almost total destruction of the UARAF's offensive capabilities. Although Egyptian losses in terms of interceptors and fighter-bombers were not as heavy as usually assessed, the attack caused such a degree of surprise and delivered such a severe blow that the leadership and entire command structure of the Egyptian military was left panic-stricken, and was unable to respond in any meaningful way for the rest of the conflict. To all intents and purposes, the outcome of the June 1967 War was decided by the IDF/AF's operations against UARAF air bases. In turn, this operation not only became a template for future military campaigns, but also strongly influenced leading air forces around the world, encouraging them to fortify their air bases and protect their aircraft with the help of hardened aircraft shelters (HAS), as well as to develop specialised runway-penetrating and runway-denial weapons.

Once it was free from the Egyptian threat, around noon on 6 June 1967 the IDF/AF launched similar operations against Jordanian and then Syrian and even one Iraqi air base. This wave of IDF/AF attacks resulted in the almost total destruction of the Royal Jordanian Air Force (RJAF), and forced the weakened Syrian Arab Air Force (SyAAF) on to the defensive not only for the rest of the war, but for several years beyond.

The neutralisation of the Arab air forces achieved by the IDF/AF in turn enabled Israeli ground forces to rapidly advance into Sinai and the West Bank, and defeat much larger enemy forces. Furthermore, it enabled the IDF to subsequently invade Syria and secure the Golan Heights that dominate northern Israel.

Historically, it has been argued that success in the 1967 war came at a heavy price in terms of aircraft losses and eventually marked the end of the French era in the IDF/AF. However, while the French government imposed an arms embargo, French companies continued to cooperate with Israel. Correspondingly, the acquisition of 51 Dassault-Breguet Mirage 5 fighter-bombers ordered before June 1967 continued and all were clandestinely delivered by 1971, despite the embargo. In order to conceal this development, the fighters were officially declared as 'manufactured by Israel Aircraft Industries' (IAI).

Meanwhile, the place of France as Israel's main supplier of weapons was soon taken by the US. Skilful manoeuvring by Israeli politicians ensured an understanding with Washington and led to several successive arms deals signed in the late 1960s and early 1970s, which included delivery of the McDonnell Douglas A-4 Skyhawk (locally named Ahit – eagle), Sikorsky S-65/CH-53 (Yasur – petrel) and the technologically advanced McDonnell Douglas F-4E Phantom II (Kurnass – sledgehammer).

1968–73 and the Yom Kippur War

While the June 1967 War had officially ended, hostilities continued and before long Egypt and Israel slid into the War of Attrition, fought between 1968 and 1970. Full-scale war broke out along the Suez Canal, consisting of daily attacks on Israeli fortifications on the eastern side of the Canal and Israeli retaliation with artillery and air strikes. The IDF/AF employed its full arsenal of fighters in these air strikes, including the old Mystère IVAs, Super Mystères and Mirage IIIs, and the newly arrived A-4 and F-4E. In an attempt to break Egypt's will to fight, but also in order to intensify the conflict with the aim of making the US more closely involved in supporting Israel, the IDF/AF eventually began targeting military bases and infrastructure targets deeper inside Egypt. Eventually, these deep strikes prompted a Soviet military intervention and the stationing of Soviet-manned MiG-21 units as well as SA-3 surface-to-air missile (SAM) sites along the Suez Canal, in addition to a vast expansion of Egyptian air defence capabilities. In a notable engagement on 30 July 1970, a well-planned IDF/AF ambush led to the destruction of five Soviet-flown MiG-21s. However, since it was unable to counter the developments in air defences, the IDF/AF ultimately suffered heavy losses in A-4s and F-4s, and Israel eventually agreed to a ceasefire. The War of Attrition ended in August 1970, with one IDF/AF commander noting that the 'anti-aircraft missile has bent the aircraft wing'. There was a widespread understanding that new tactics and weapons had to be developed to fight – and win – against the Soviet-designed integrated air defence system.

Meanwhile, smaller-scale operations were launched by the IDF along ceasefire lines to Jordan, Syria and into Lebanon, primarily against the camps of Palestinian insurgents. The IDF/AF supported such operations as well, primarily deploying fixed-wing aircraft (A-4s, Super Mystères and F-4s) to destroy Jordanian and Syrian artillery and armoured fighting vehicles (AFVs), while rotary-wing S-58s and the new Bell UH-1 were used to pursue and capture terrorists infiltrating Israeli territory. While a tentative ceasefire was agreed with Jordan in 1971, tensions and exchanges of blows between Israel and Syria continued well into 1973.

On 6 October 1973, Egypt and Syria took Israel by surprise by launching a concerted attack on Yom Kippur, the holiest day in the Jewish calendar. Egyptian forces crossed the Suez Canal and captured the so-called 'Bar Lev Line', a line of fortifications along the eastern side of the Canal. Within the following two days, Egyptian forces established several bridgeheads as deep as 15km (9.3 miles) along the eastern side of the water obstacle. Aiming to retake the Golan Heights, Syrian forces launched a massive armoured onslaught that penetrated Israeli positions in the south and almost resulted in the capture of crucial bridges on the River Jordan.

Israeli security doctrine has long been based on small regular forces and large reserve forces. In 1973, the gap between the beginning of hostilities and the mobilisation of major reserves was filled by the regular forces – and the IDF/AF. In this scenario, the air force's main mission was to prevent the advance of enemy forces at any cost. It took Israel three days to mobilise and deploy its reserve forces and during these days the IDF/AF suffered extensive, often painful losses, especially to the new types of SAMs. However, after these three days Egypt stopped its attacks, while the Syrian forces were exhausted and had suffered extensive losses in the face of fierce

Israeli resistance. The initiative gradually passed to the Israeli side, which the IDF ably exploited in order to prepare for major counterattacks on both fronts.

Indeed, although battered by the losses sustained at the beginning of the war, the IDF/AF was to lead these counteroffensives, firstly by neutralising Syrian air defences along the front line by striking targets deeper inside that country, and then by establishing air superiority over the battlefield and immediately behind, by bombing SyAAF air bases and engaging Syrian interceptors in numerous air combats. In this way, the IDF/AF enabled the Israeli ground forces to advance even deeper into Syria than had been the case during the June 1967 War.

On the Egyptian front, the IDF/AF remained unable to inflict attrition upon the Egyptian Air Force until Israeli ground forces crossed the Suez Canal and destroyed a number of SAM sites, thus creating a 'gap' within the Egyptian air defences. Once this gap was established, the IDF/AF was able to operate more freely, and cause heavy losses not only to the EAF, but also to Egyptian ground forces.

The war on the Egyptian front ended after some of the most intensive air warfare since World War II and only with a second UN-mediated ceasefire, on 24 October 1973. Fighting on the Syrian front actually came to an end only in May 1974.

Overall, the IDF/AF ended this conflict claiming the destruction of 370 Arab aircraft and 55 helicopters in the air and 103 on the ground, in exchange for an officially admitted loss of 107 of its own aircraft and helicopters.

1974–89: no target too far

The years after the 1973 conflict saw a period of unprecedented growth and development of the IDF/AF. Understanding that the IDF/AF was the only force standing between Israel and defeat in the first days of a possible major war in the future, Israel strengthened its air force, drawing upon lessons learned from the 1973 conflict. The IDF/AF thus saw the introduction of vastly superior interceptors (in the form of the McDonnell Douglas F-15 Eagle, locally named Baz – buzzard); of a very potent antitank capability (in the form of the Hughes Model 500 Defender, named Lahatut – magic trick, and then the Bell AH-1 Cobra, named Tzefa – viper); and the development and service entry of the locally manufactured IAI Kfir (lion cub) fighter-bomber (developed with US and French assistance, and partially manufactured from components delivered from France). Finally, in 1979 Israel began the process of introducing the General Dynamics F-16A/B Fighting Falcon (named Netz – sparrowhawk) as replacement for the ageing fighter-bombers of French origin.

Correspondingly, the early 1980s saw the IDF/AF beginning to demonstrate its new might and confidence. In 1981, the air force claimed its first kill against the Soviet-built MiG-25 fighters capable of reaching speeds of up to Mach 3; the IDF/AF also launched a spectacular attack on the construction site of Iraqi nuclear reactors outside Baghdad. Only a year later, the IDF/AF convincingly put to the test all the lessons learned from experiences during the October 1973 War, in the course of the Israeli invasion of Lebanon. Specifically, on the third day of that war, the air force attacked and largely neutralised the Syrian integrated air defence system located in the Bekaa Valley of eastern Lebanon. When the SyAAF reacted by scrambling dozens of its interceptors, and also with attempts to hit advancing Israeli ground forces with its own fighter-bombers, the IDF/AF engaged these under particularly challenging, yet pre-planned conditions, and claimed 82 Syrian aircraft shot down. All of this was achieved with only minimal losses to the IDF/AF.

From 1982 onwards the IDF/AF continued its involvement in the war in Lebanon, primarily in the form of ground attacks against Palestinian and other militant organisations, and in occasional dogfights with Syrian fighters (the last of these took place in 1985, pitting F-15s and F-4s against MiG-23s, and resulted in claims for two Syrian fighters shot down).

Another long-range attack was launched on 1 October 1985, when F-15s flew 2,060km (1,280 miles) to attack the PLO Headquarters in Tunis, the Tunisian capital, spectacularly demonstrating the new capabilities of 'Israel's long arm', as the IDF/AF was often termed in the media. During the mid-1980s the IDF/AF began adding the advanced F-16C/D Barak (lightning) and the upgraded F-4E Kurnass 2000 to its arsenal, but was simultaneously forced to abandon the development of an ambitious attempt to develop a world-class, low-cost and highly manoeuvrable lightweight fighter – the IAI Lavi (young lion) – with help from the US, for financial reasons. Although the first Lavi prototype flew on 31 December 1986, production costs spiralled out of control due to a decision to purchase only 75 airframes instead of the promised 150, and US refusal to finance what could become a direct competitor to its F-16 on the international market. The cancellation of the Lavi, announced on 30 August 1987, resulted in some 6,000 IAI employees losing their jobs, and was felt strongly in Israel for some years to follow. Nevertheless, the project yielded significant benefits for the country and the IDF/AF, since it resulted in the transfer of technology and know-how from the US to Israel, and led to IAI and companies like Rafael, Elisra and Elbit becoming world-class producers of radars, electronic countermeasures (ECM), communication systems, missiles and precision-guided munitions, as well as unmanned aerial vehicles (UAVs).

1990–2013: modernisation and new threats

The last 20 years of the Israeli Air Force were characterised by gradual modernisation of its fleet and its adaptation to newly emerging challenges. On the one hand the IDF/AF was forced to develop new capabilities to counter asymmetric warfare as well as to fly extremely precise attacks against militants hidden among civilian populations. On the other hand it had to develop the capability to fly large-scale missions against heavily protected and hardened targets thousands of kilometres from Israel – a result of the perceived threat from Iran. In recent years these two threats have come to be considered of equal importance, as evidenced by Palestinian rocket attacks from the Gaza Strip and Hezbollah attacks from southern Lebanon, some of which developed into a strategic threat, as seen during the Second Lebanon War in 2006.

Correspondingly, the force was re-equipped with some of the latest and most modern Western hardware, starting with the McDonnell Douglas AH-64A Apache (named Peten – python) in 1990 and Boeing AH-64D Longbow Apache (Saraf – serpent) in 2005. It modernised ageing CH-53 helicopters to the Yasur 2000 standard in 1993, and purchased Sikorsky UH-60 Black Hawk (Yanshuf – owl) helicopters in 1994. In 1998, the air arm started the acquisition process for the Boeing F-15I (Ra'am – thunder), followed by the Lockheed Martin F-16I (Sufa – storm) in 2004, and the Gulfstream V Nachshon (pioneer) Shavit (comet) intelligence-gathering platform and the Gulfstream 550 Nachshon Eitam (fish eagle) airborne early warning and control aircraft in 2006. From 2007, the air arm began receiving advanced UAVs including the Elbit Hermes, IAI Heron (Shoval – trail) and IAI Heron TP (Eitan – firm), all of which represent a significant new enhancement in capability. Each of these platforms fulfils specific functions within the Israeli Air Force order of battle, with the aim of tackling Israel's future military challenges as outlined above.

The future IASF

In 2013, as this book was being prepared, the Israel Defense Forces announced a potentially far-reaching programme of restructuring, including reductions in terms of equipment and personnel, the latter described by one insider as 'the most radical downsize since 1973'. At the same time, however, the IASF is preparing to receive new aircraft either already on order, or likely to be acquired in the coming years.

In a concerted effort to modernise the IASF fleet, orders have been placed for the Lockheed Martin C-130J Hercules (Shimshon – Samson), Lockheed Martin F-35 Lightning II (Adir – mighty) and Alenia Aermacchi M-346 Master (Lavi – young lion). Meanwhile, the rotary-wing fleet is likely to be bolstered in the future by the Bell Boeing V-22 Osprey tiltrotor transport. Looking further ahead, and the latest Sikorsky CH-53K variant or the Boeing CH-47 Chinook are likely candidates to replace the CH-53 within the next 10 years. Urgently required is a new aerial refuelling tanker, to replace the Boeing 707 fleet. Washington has offered to donate surplus US Air Force Boeing KC-135 Stratotankers, although the latter deal may hinge upon renewed Israeli peace talks with the Palestinians.

Importantly, when it comes to future arms transfers from the US, Washington is likely to combine the supply of equipment to the IDF within the framework of broader deals covering other Middle East clients, and in particular Saudi Arabia and the United Arab Emirates. The US announced a Middle East arms deal worth USD 10 billion in July 2013, which could include the long-expected V-22s and KC-135s as well as new fighter radars and long-range anti-radar missiles. The latest policy from the US aims to preserve the qualitative edge of the IDF while ensuring that the more hawkish Israeli policy-makers exercise restraint.

The budget cuts announced in 2013 have not hit the IASF as hard as some observers expected, but the downsizing in the AH-1 force with emphasis on the training of new attack helicopter pilots in the Air Force Academy, and the consolidation of the last two F-16A/B squadrons may just mark the beginning of a wave of future force reductions. Also subject to recent change is the C-130 force, with the consolidation of the existing two squadrons. Notable, however, is the fact that changes to both the F-16A/B and C-130E/H communities also serve to prepare for the arrival of the new F-35 and C-130J.

Chapter 2

IASF COMBAT OPERATIONS IN THE 21st CENTURY

Since the beginning of the new millennium Israel has experienced a period of continuous conflicts with some of its neighbours. The Hezbollah militia (designated a terrorist organisation by Israel, US and – more recently – the European Union) in southern Lebanon (which was vacated by Israel in 2000), and militant Palestinian groups, foremost Hamas, operating from the Gaza Strip (vacated by Israel in 2006), have challenged Israel on numerous occasions during this period, and the IASF has been deployed in combat against them on a number of occasions.

The IASF was the arm chosen by the Israeli government to engage other threats, too, including the suspected North Korean-built nuclear reactor in Syria, in 2007; armament convoys en route from Iran to the Gaza Strip in Sudan in 2009, 2010 and 2011; and targeted killings of persons in charge of various militant organisations or in illicit activities endangering the security of Israel. Although never officially confirmed by Israeli authorities, it is obvious that the IASF has also exploited opportunities that have presented themselves due to the civil war raging in Syria since 2011. These missions have targeted specific weapons-storage facilities in Syria, as well as convoys carrying arms for Hezbollah. Finally, the IASF has been involved in enforcing a maritime blockade of the Gaza Strip and the capture of several ships carrying arms to militant Palestinian groups in Gaza, including rockets, firearms and ammunition.

Operation Defensive Shield, 2002

Following the failure of Arab-Israeli peace negotiations, in March 2002 a series of 11 suicide attacks hit Israeli cities and Israeli military units deployed in the West Bank, killing 81 civilians and 30 soldiers. In response, the Israeli government launched Operation Defensive Shield. This began with the mobilisation of 20,000 reserve soldiers and an advance into areas controlled by the Palestinian Authority in an attempt to destroy the strongholds of Palestinian militants, and capture or destroy their weapons, facilities and bomb-manufacturing facilities.

The IASF supported the operation in many ways: fighter aircraft initiated air-to-ground strikes, attack helicopters provided close air support to infantry forces fighting on the ground, and intelligence-gathering and control missions involved the UAV fleet. From a military point of view, Defensive Shield was a major success. Israel managed to put an end to the wave of terror and to return confidence and security to the state of Israel, within a short period of time. Meanwhile, Israeli military casualties remained relatively limited.

Ebb and Flow and escalation in the Gaza Strip, 2006

In reaction to the failure of Arab-Israeli peace negotiations, the Palestinians in the Gaza Strip also launched an uprising, in September 2000, resulting in widespread protesting, rioting and attacks on the Israeli military and civilians. By July 2006, Israel suffered 7,918 injured (including 5,521 civilians) and 1,109 killed (including 783 civilians). In an attempt to end the violence and draw future Israeli borders, the government then decided to unilaterally withdraw from the Gaza Strip. In September 2005, all IDF forces and 8,000 civilians were evacuated from their bases and settlements in this area.

AH-1F serial number 355 takes off from Palmachim, armed with two pairs of TOW missiles. (Ofer Zidon)

UH-60 helicopters from No. 123 'Desert Owls' Squadron await the embarkation of infantry. (IDF Spokesperson)

Unfortunately, the Palestinian response to the Israeli move was to launch a terror campaign from the Gaza Strip. The Palestinian militia used the ruins of Israeli towns in Gaza to launch unguided rockets towards Israeli towns on the Israeli side of the border. The number of rocket launches grew steadily, reaching a total of 130 in January 2006.

Israel's response was based on the massive artillery shelling of suspected rocket launch sites, round-the-clock UAV and other intelligence monitoring of the Gaza Strip as part of a real-time search for rocket-launch teams, and 'pinpoint' attacks against terror leaders. The concerted Israeli counteroffensive against the Palestinian terrorists succeeded in temporarily reducing rocket launches to 70 during April 2006.

Operation Summer Rain, 2006

Operation Summer Rain commenced on the morning of 28 June 2006, after Palestinian militants captured IDF Corporal Gilad Shalit from a military outpost on the Israeli side of the Gaza ceasefire line. Israeli tanks, AFVs and troops entered the Gaza Strip, just hours after the IASF destroyed two major bridges and the main power station in Gaza, effectively shutting down the electricity and water supply. Attack helicopters supported IDF infantry and armour incursions into the Gaza Strip, and Israeli jets created sonic booms. Because of links between the Syrian government and the leadership of Hamas, and considering Syria responsible for the capture, the IASF was also deployed to signal a message to Damascus. At 04.00 on the morning of 28 June 2006, a formation of F-16s over-flew the residency of Syrian President Bashar al-Assad in Latakia. The message remained ignored and Shalit remained imprisoned in Gaza until his release in October 2011.

Operation Change of Direction: Second Lebanon War, 2006

In May 2000, after nearly 20 years of military occupation of southern Lebanon, and in the face of increasing pressure from Hezbollah, Israel felt compelled to withdraw from that country. After the withdrawal, the Iranian-backed Hezbollah deployed its forces along the Israeli-Lebanese border. With the IDF focused on its June 2006 campaign in the Gaza Strip, a Hezbollah force crossed the Israel-Lebanon border on 12 July 2006, setting an ambush for an Israeli border patrol. Three IDF soldiers were killed and two were taken prisoner. That night, Israel launched a massive bombing campaign against Lebanese infrastructure and Hezbollah installations, in the process knocking out the entire Hezbollah arsenal of medium-range surface-to-surface missiles. Hezbollah responded to the Israeli air offensive with a concerted series of rocket launches against Israeli civilian targets, mainly against towns in northern Israel.

Israel's objectives in the Lebanon campaign were similar to those in the Gaza campaign: the unconditional return of imprisoned soldiers and a permanent removal of Hezbollah as a viable fighting force in southern Lebanon. Initially, it was hoped that air power alone would achieve these objectives, but Israel's defences against the

Modern Israeli Air Power

An F-16A Netz from No. 140 'Golden Eagle' Squadron armed with four M117 general-purpose bombs.
(Nevatim Archive via Ofer Zidon)

Hezbollah rocket threat were far from perfect. The IASF tracked down and destroyed Hezbollah's medium-range artillery rockets within the first hour of the war. During the campaign's first seven days, the IASF flew some 2,000 fighter and attack helicopter sorties. Very often, there were 40 to 70 IASF aircraft operating simultaneously in the compact battlespace above southern Lebanon. Gulfstream Nachshon surveillance aircraft and UAVs searched for targets, while fighter-bombers were deployed to hit targets from stand-off ranges, usually conducted at altitudes above 12,000ft (3,658m) to keep their crews safely beyond Hezbollah anti-aircraft fire. Attack helicopters conducted nap-of-the-earth operations. Accordingly, there was a continuous hail of fire falling through skies full of aircraft, making flawless time and space deconfliction an

The AH-64D Saraf played the crucial role in on-call CAS operations over southern Lebanon, in 2006. This example was photographed while preparing for take off from Ramon.
(Ofer Zidon)

Chapter 2

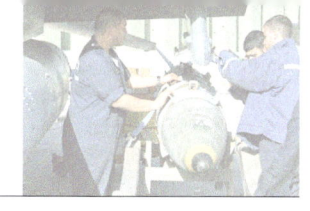

An F-15I Ra'am in 'heavy' air-to-ground configuration prepares to leave its HAS prior to an attack mission on targets in Lebanon. (Ofer B.)

ever-present airspace management requirement. The IASF met all these requirements with resounding success. However, it lacked the ability to intercept rockets once these were fired, or to destroy small launchers, even when supported by artillery. With a daily average of 100 rockets launched against Israel, a different approach was needed. Fearing heavy casualties if they were to employ a fully developed contingency plan for a joint air-ground counteroffensive designed for just such a possible circumstance, and knowing that IDF ground forces lacked the necessary readiness, the Israeli political and military leadership opted for a limited, rolling invasion of Lebanon, launched on 17 July 2006, with the objective of securing major Hezbollah strongpoints along the Israeli-Lebanese border.

By 6 August 2006 eight IDF brigades (three armoured and five infantry) were operating in southern Lebanon, with the objective of destroying Hezbollah infrastructure and hunting down rocket launch teams. However, Hezbollah remained capable of firing an average of 100 rockets per day. On 11 August 2006 the United Nations Security Council adopted Resolution 1701 calling for Israel and Lebanon to support a permanent ceasefire and a long-term solution based on the following principles and elements: full respect for the Blue Line (the ceasefire line and de-facto international Israel-Lebanon border) by both parties; the establishment of an area free of any armed personnel and weapons other than those of the government of Lebanon and of UNIFIL (United Nations Interim Force in Lebanon) between the border and the Litani River; disarmament of all armed groups in Lebanon.

After three weeks of strenuously resisting a ground offensive, the IDF finally bowed to the calls for a ground invasion. It tripled the number of units deployed inside Lebanon during the final 72 hours of combat, and launched a large-scale heliborne operation. Due to poor coordination among ground-force elements and much heavier Hezbollah resistance than expected, heavy casualties became unavoidable and an IASF CH-53 was shot down with the loss of five aircrew on the evening of 12 August 2006, only minutes after disembarking dozens of troops.

Modern Israeli Air Power

For the most part, the IASF performed to its usual high standards of competence, and registered impressive achievements, especially in regards to pre-emptive attacks on Hezbollah's medium-range rockets during the opening night, and highly effective time-sensitive targeting operations during the following weeks. It flew more than 15,000 sorties including 12,000 fighter-bomber sorties, 3,000 rotary-wing sorties and 2,500 transport sorties to counter Hezbollah's rocket attacks and support IDF ground forces. The IASF attacked 7,000 targets in Lebanon, at an average rate of 340 sorties a day, providing abundant on-call close air support, but also promptly evacuating many wounded IDF troops by UH-60 helicopters under heavy fire.

Threats from Iranian and Syrian nuclear ambitions, 2007

On 6 September 2007, Syria stated that it had detected Israeli combat aircraft penetrating deep into its territory the previous night. According to Syria, its air defences opened fire on these aircraft, forcing them to jettison their drop tanks and ordnance. Later that day Turkey reported that it had found external fuel tanks from an F-15 not far from the Turkey-Syria border. As in the case of numerous past incidents, Israel officially ignored the issue. Foreign media reports suggested that the target of the alleged attack was an installation west of the city of Dayr az-Zawr in eastern Syria. Based on satellite images, some Western observers claimed that the site was some kind of nuclear installation, perhaps built with North Korean support.

Over the years, together with increased threats to Israel's security from 'third-circle' nations like Iran, the IASF has developed its power-projection capabilities, based on the long-range, all-weather, precision-strike F-15I fleet, supported by the newer F-16I

One of the Ramat David-based F-16 Baraks armed with a JDAM.
(IDF Spokesperson via Ofer Zidon)

Chapter 2

An AH-64A Peten from No. 190 'Magic Touch' Squadron on its way to Gaza.
(Ofer Zidon)

An F-15I in air-to-ground configuration is inspected by ground crew before leaving its HAS.
(Ramon Archive via Ofer Zidon)

25

Modern Israeli Air Power

An F-16 armed with two 500lb (227kg) GBU-32 JDAMs. Due to proximity of the battlefield, F-16Cs were able to fly combat sorties without carrying any drop tanks.
(Ofer Zidon)

fleet. Both types can employ conformal fuel tank (CFT) packs, extensive all-weather navigation and targeting systems, and precise weapons including laser- and GPS-guided munitions to reach distant objectives and attack these with a high probability of success.

The main component in such an attack would likely require top cover from F-15 fighters; command and control and communication relay using the Gulfstream 550 Conformal Airborne Early Warning (CAEW) system; and in-flight refuelling from the Boeing 707. Another component of such a mission would be a long-range combat search and rescue (CSAR) element circling at a safe distance and waiting to be activated if needed. A typical CSAR component is based on both types of IASF assault helicopter, the CH-53 and UH-60, accompanied by the C-130 for aerial refuelling.

Armed with GBU-10 laser-guided bombs, an F-15I from No. 69 'Hammers' Squadron prepares to taxi from its HAS to the Hatzerim runway.
(Ofer B.)

Chapter 2

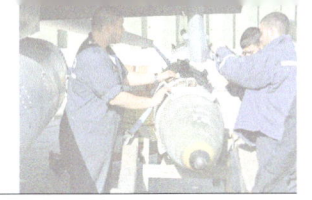

Operation Cast Lead, December 2008 to January 2009

The Gaza Strip continued to be a source of rocket attacks and mortar shelling against Israeli towns. Rocket attacks by various Palestinian factions from Gaza gradually became more intense and more accurate thanks to the development of heavier, longer-range and more accurate rockets. Larger numbers of Israeli civilians found themselves under fire to the point where the Israeli leadership saw these attacks as a strategic threat to Israel. A decision was made to launch a massive operation against the Gaza strip in order to regain deterrence and put an end to the rocket attacks.

Operation Cast Lead commenced at noon on 27 December 2008, the IASF first bombing 50 infrastructure targets including Hamas headquarters, weapons storage facilities and training camps with devastating success. The second wave hit all the known rocket launch sites, attempting to suppress a possible counterattack. The IASF used smaller GPS-guided bombs when the target was in close proximity to civilian buildings, to minimise collateral damage.

The IASF bore the brunt of fighting around the clock, constantly monitoring the Gaza Strip using its UAV fleet and blimps equipped for surveillance, and flying air strikes – both planned attacks against known targets, and real-time responses to targets recently uncovered by intelligence.

The IASF also used its fighter-bombers to destroy tunnels built by the Palestinian militia under the Gaza-Egypt border. The tunnels were used as supply lines to Gaza and as weapon storage facilities. Further air attacks were carried out against infrastructure targets including weapons manufacturing and storage facilities, rocket launch sites, launch teams, launchers and trucks transporting unguided rockets. Another component of IASF operations included tracking down leaders of various Palestinian militant groups, launching air strikes on the basis of near-real-time intelligence.

On 17 January 2009 Israel surprisingly announced a unilateral ceasefire, conditional on the conclusion of further rocket- and mortar attacks from Gaza, and began withdrawing its forces. The last IDF ground troops vacated the area three days later, although the Palestinians fired 17 additional rockets in the meantime.

An AH-64D Saraf from No. 113 'Hornet' Squadron at Ramon, armed with Hellfire missiles. (Ofer Zidon)

Modern Israeli Air Power

Alleged operations in Sudan, 2009 to 2012

According to reports published in Israeli and foreign media, the IASF launched a number of attacks on selected targets inside Sudan several times since early 2009. The first of these was reportedly related to Operation Cast Lead, and was undertaken in January 2009. According to Sudanese officials, it hit a convoy of 16 trucks under way in the desert east of Port Sudan. Loaded with refugees and light weapons, these vehicles were apparently secured by 39 armed persons. Supposedly, the convoy was completely destroyed. The second such attack seems to have occurred on 11 February 2009 and hit a convoy of 18 trucks that, according to Sudanese officials, was 'loaded with up to 1,000 civilians about to be smuggled over the border to Egypt'. According to the same source, 119 persons were killed, including 56 smugglers and 63 smuggled persons of not only Sudanese, but also Eritrean, Ethiopian and Somali nationality. While US sources reported that the attacks in question hit convoys carrying arms smuggled from Iran for Hamas in the West Bank, Israeli officials refused to offer any commentary, and unofficial Israeli sources indicated the deployment of armed UAVs in at least one of these two operations.

An AH-64D-I parked on the No. 113 'Hornet' Squadron apron. The helicopter is armed with Hellfire missiles. (IAF Magazine)

Chapter 2

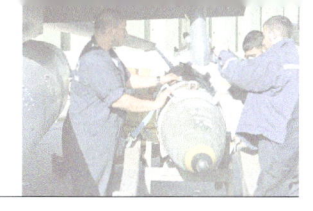

F-16I tail number 854 from No. 201 'The One' Squadron taxis to its HAS, armed with a Rafael Spice guided ammunition, Python 5 and AIM-120.
(Ofer Zidon)

In June 2009, the IASF hit a convoy carrying weapons while this was approaching Sudan's border with Egypt. The attack was subsequently confirmed by Israeli officials, as was another such operation, launched in April 2011, when an air strike apparently delivered by an IASF UAV hit a car under way outside Port Sudan, killing two persons, including one leading Hamas official.

Finally, in October 2012, after a series of massive explosions destroyed a part of the Yarmouk complex of military facilities near Khartoum, the Sudanese government accused Israel of another attack, and some Western observers claimed that the IASF had destroyed a suspected storage site of Iranian-made surface-to-surface missiles. However, no conclusive evidence was ever delivered for any kind of Israeli involvement in this affair.

The F-16I entered service just in time to fly its first combat sorties during the Operation Cast Lead. This example from No. 253 'Negev' Squadron, armed with a GBU-10 laser-guided bomb and an AIM-9 Sidewinder, was photographed while taxying to the runway at Ramon.
(Ramon Archive via Ofer Zidon)

Modern Israeli Air Power

Iron Dome – first deployment, March 2012

On 11 March 2012 the IDF killed the Secretary General of the Popular Resistance Committees in Gaza, Zuhair al-Casey, as he was in the final stages of preparing another attack against Israelis in Sinai. This was to follow his success in killing eight Israeli civilians and injuring over 40 in an attack against a bus and private vehicles on their way to Eilat six months earlier. In retaliation for the targeted killings, Islamic Jihad terrorists fired over 200 rockets against civilian targets in the major cities in the south of Israel. The rockets fired included locally made Kassams and military-standard Grads, smuggled into Gaza from Egypt. During this period the IASF attacked terrorist targets, including launch sites, rocket manufacturing and storage facilities and also directly hit terrorists while in the process of launching rockets.

The main defence system now in use against these rockets is the Iron Dome. Rafael developed this air defence system during the previous six years, as the main lesson from the Hezbollah rocket attacks on Israel during the 2006 Second Lebanon War.

Iron Dome is a mobile air defence system designed to operate against short- to medium-range rockets with range of 4 to 70km (2.5 to 43.5 miles). It saw full-scale operation for the first time in March 2012, with three batteries deployed around the Gaza Strip to defend the major cities in southern Israel: Beer Sheba, Ashkelon, Ashdod, Ofakim and Netivot, with a combined population of over 1 million.

Iron Dome works in three phases. First, the system's radar detects the rocket launch and then the system's control centre calculates the incoming rocket's trajectory, and it then decides whether to intercept if the projected area of impact is populated. The third stage is the launch of the interceptor, which destroys the incoming rocket after detonation by a proximity fuse.

During this four-day conflict, Iron Dome detected 68 rockets of the 200 launched as potential threats to populated areas and successfully intercepted 55 of them. The interception success rate exceeded 80 per cent.

An Iron Dome interceptor missile departs its launcher. Rafael has begun a major effort to reduce the price of each interceptor to a 'a few thousand dollars', using simpler and cheaper materials and subsystems.
(Nechemia Gershuni)

Chapter 2

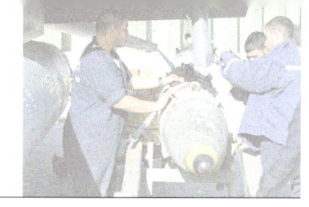

Operation Pillar of Defense, 2012

This operation was launched on 14 November 2012 with the killing of Ahmed Jabari, second-in-command of Hamas. The operation came as the response to a massive increase in rocket launches from Gaza – up to 100 rockets in a 24-hour period prior to the operation – an attack on an Israeli patrol vehicle, and a tunnel explosion near a group of Israeli soldiers, the latter two incidents occurring on the Israeli side of the ceasefire line.

Operation Pillar of Defense was terminated on 21 November 2012, after a ceasefire was announced, following negotiations between Hamas and Israel, and mediated by Egypt. The IDF assessed that its operations resulted in hits on more than 850 different targets, and delivered significant damage to the command and control apparatus of Hamas, successfully targeting seven of its top commanders and heavily damaging several facilities and military bases, as well as dozens of tunnels.

Among the Hamas capabilities that the IDF assessed as damaged were dozens of long-range (over 40km/ 24.8 miles) and hundreds of short- and medium-range rocket launchers.

Surveillance blimps, like this example manufactured by Top Eye View, Inc., are playing an increasingly important role in IASF operations. (Top Eye)

Modern Israeli Air Power

In terms of the defence of Israel and its population, arguably, the number of rockets fired by different Palestinian groups from within the Gaza Strip between 14 and 21 November 2012 did not significantly decrease. The Palestinians launched a total of 1,506 rockets in seven days, of which 875 were assessed as headed for empty areas, and 58 for urban areas. Although failing to intercept 152 different rockets, the Iron Dome batteries (each consisting of three launchers with 20 missiles each) successfully intercepted 421 projectiles. Since each Iron Dome missile costs around USD 50,000 and two are usually expended in the interception of any rocket headed towards an urban area, this led to some discussion concerning the economic aspects of fielding the system. However, it is undeniable that overall the Iron Dome system achieved an impressive success rate of 75 per cent.

Air strike against convoy in Syria, 2013

On 31 January 2013 the IASF launched an attack against a convoy alleged to be carrying advanced anti-aircraft missile systems from Syria to Hezbollah in Lebanon. The air strike occurred northwest of Damascus, not far from the border with Lebanon. Foreign sources alleged that at the time of the attack, the convoy had stopped inside the Syrian Scientific Studies and Research Centre, Syria's main research centre for biological and chemical weapons.

This was the first time that Israel had attacked Syria since the supposed raid on an alleged nuclear installation in 2007. Israel did not take direct responsibility for the bombing, but Israeli Defence Minister Ehud Barak hinted that it was behind the attack. In the past, Israel had announced that it would not let strategic weapons systems (including advanced anti-aircraft missiles and chemical and biological weapons) make their way from Syria into the hands of Hezbollah in Lebanon.

An F-15D Baz from No. 133 'Twin Tail' Squadron at Tel Nof undergoes inspection before leaving its shelter for a ground-attack mission. It is armed with JDAM guided bombs and Rafael Python AAMs.
(*IAF Magazine*)

INTERNATIONAL ACTIVITIES IN THE 21st CENTURY

International exercises

For many reasons, training abroad is of critical importance to the IASF. Training areas in Israel are small, local airspace is crowded with civil aviation movements, there are numerous restrictions on training over populated areas, and Israel has reliable weather for all but two months of the year. These factors combine to offer very limited training scenarios, with which aircrews quickly become familiar.

In recent years IASF squadrons and their various aircraft types have increasingly participated in exercises and deployments abroad. F-15I, F-16I and F-16C/D squadrons have participated in Exercise Red Flag in the US, with a deployment in 1999 followed by participation in the Canadian Maple Flag exercise in 2005 (F-16C/D). Since 2003, F-15 Baz and F-16 Netz and Barak squadrons have regularly deployed to Decimomannu in Sardinia, where they train with the Italian Air Force, making use of large training areas and unfamiliar terrain. Between 2004 and 2008, when Israeli relations with Turkey deteriorated, Israeli deployed F-15s and F-16s to Turkey for the Anatolian Eagle exercise. In 2013, F-16C/D squadrons conducted a deployment to Bulgaria.

Beginning in 2004, C-130 units have deployed to Romania, most recently in 2011, for low-level training in the Carpathian Mountains. CH-53s deployed to the same area, practicing high-altitude operations over mountainous regions and in bad weather. The IASF's other helicopter squadrons, equipped with the UH-60 and AH-64 Peten and Saraf, have deployed to Greece since 2010, training in long-range search and rescue missions with the Hellenic Air Force in unfamiliar areas and weather conditions and improving cooperation. Even the ferry flights required for such deployments are considered a training mission, giving crews the opportunity to practice the planning and execution of long-range flights, navigation over unfamiliar areas and aerial refuelling.

Foreign air force units also deploy to Israel, with US Air Force squadrons in particular visiting Nevatim on a regular basis as part of a bilateral defence agreement. Noteworthy is the annual Juniper Stallion series of exercises held between the European Command and the IDF. This approximately two-week exercises involves bilateral F-15/F-16 air-to-air training missions designed to improve interoperability and cooperation between the US and Israeli air arms. Italian, Romanian, Polish and Greek air force units have also deployed to Ovda to conduct joint exercises with the IASF. A new partner for international exercises is France, and in 2013 the French Air Force was invited to train in Israel and, in return, IASF pilots were invited to fly in France for the first time. Future cooperation with Italy is likely to involve training exercises to coincide with the IASF's acquisition of the C-130J.

Modern Israeli Air Power

F-16I serial number 407 from No. 253 'Negev' Squadron is seen during a deployment to Greece by the Sufa squadrons in October 2012.
(P. Panagiotopoulos/ICARUS)

An F-16C from USAFE's 31st Fighter Wing at Aviano AB touches down at Nevatim during a US deployment to the base in 2006.
(Ofer Zidon)

Chapter 3

Serial number 77-089, an F-15A assigned to the USAF's 102nd Fighter Wing, a unit of the Massachusetts Air National Guard, gets airborne from Nevatim during a Juniper Stallion training deployment to the base in 2006.
(Ofer Zidon)

A E-3 Sentry AWACS lands at Nevatim during the USAF deployment to the base in May 2006.
(Ofer Zidon)

Modern Israeli Air Power

F-16A serial number 281 from No. 115 'Flying Dragon' Squadron leads a pair of Polish F-16Cs over the Negev Desert in southern Israel.
(Maj. Roy B.)

F-15I serial numbers 238 and 234 are seen with a USAF F-15C during a Juniper Stallion deployment to Nevatim in 2006.
(Ofer Zidon)

An Italian Air Force Tornado takes off during its 2013 deployment to Ovda. In the background, a pair of F-15C Baz from the 'Twin Tail' Squadron await clearance for take-off. (Ofer Zidon)

Foreign aircrews testify that training in Israel – usually at the Advanced Training Center in Ovda – has many advantages: year-round good weather, close proximity to training areas (only a few minutes' flying time), obstacle-free low-level flying conditions, and the invaluable services given by the ATC's adversary squadron, No. 115 'Flying Dragon' Squadron. Unofficially known as the 'Red Squadron', this latter is a very efficient and well-trained unit that utilises enemy tactics against the training forces. Moreover, the IASF offers considerable experience in the organisation of deployments and exercises and the planning of missions of all sizes and for all platforms.

Humanitarian missions

Delivery of humanitarian aid to disaster areas, such as earthquakes and forest fires, is not a well-known IASF activity, but has been part of the air force's mission for many years. Its origins lie in Israel's sensitivity to the suffering of unfortunate people all over the world, following the sad circumstances of the Jewish people in World War II. Israel is also committed to supporting Jews in distress anywhere in the world. Since the mid-1980s, Israeli Air Force transport aircraft have flown to Africa, the Americas, Asia and Europe to support people in need, in response to both natural and man-made disasters. Time and again, the aircrews of Israeli Air Force C-130 and Boeing 707 transport aircraft have found themselves flying into areas with shattered facilities, bringing much needed supplies and medical assistance. During such missions the air force works closely with the Home Front Command's Search and Rescue Unit, a highly specialised organisation with expertise in search and rescue, recovery of people from collapsed buildings, and rescue in areas affected by earthquakes, floods and fires. Another important partner in these missions is the IDF Medical Corps, which is involved in establishing field hospitals and providing medical treatment.

Modern Israeli Air Power

An IDF/AF 707 flew 20 tons of humanitarian supplies to aid Muslim refugees in Bosnia on 25 July 1995. The Israeli Boeing was adorned with the title 'The Joint Israeli-Jordanian Mission to Bosnia' to indicate cooperation with Jordan, which dispatched a Lockheed L-1011 TriStar with 35 tons of humanitarian supplies. The two jet freighters flew together and unloaded their cargos side by side.
(IDF Spokesperson)

The following lists the major humanitarian aid missions conducted by the Israeli Air Force during the last decade:
- On 7 January 2005, Israel responded to Sri Lankan and Indonesian requests for aid after a tsunami hit these countries, killing over 250,000 people and leaving millions homeless. Israel sent 82 tons of medical and humanitarian aid to the affected regions, including 10,000 blankets, 12 tons of food, 17.5 tons of baby food, over nine tons of medicine and additional supplies such as generators, tents, beds and mattresses.
- On 7 September 2005, after Hurricane Katrina hit the city of New Orleans and its suburbs, leaving hundreds of thousands of people homeless, an IDF delegation left for New Orleans, taking with them 80 tons of donated tents, folding beds, bottled water, bed linen, blankets, dried food and other equipment.
- On 24 January 2006, following the collapse of a building in Nairobi, Kenya and an appeal made by Kenyan Vice President Moody Awori to Foreign Minister Tzipi Livni, Israel responded by sending the Home Front Command's Search and Rescue Unit to rescue victims trapped under the rubble.
- In January 2010 Israel provided a field hospital with a crew of 220 medical experts after an earthquake struck Haiti. The specialist search and rescue service recovered survivors from inside collapsed buildings.
- On 25 March 2011, following the tsunami that hit Japan, Israel was one of the first countries to send aid to the area, providing a field hospital and other emergency supplies.

The Israeli Air Force has manifested its commitment to help Jews in distress on several occasions:
- On 24–25 May 1991 some 14,500 Ethiopian Jews were brought to Israel over two days using an airlift of IDF/AF C-130 and Boeing 707 aircraft and airliners operated by civilian carrier EL AL. Six Boeing 707s conducted 12 sorties, bringing some 5,600 people to Israel.
- On 19 April 1994, following the bombing of the Jewish Community Centre in Buenos Aires, Israel sent a Boeing 707 carrying a search and rescue team and 20 tons of supplies.

Chapter 4

IASF AIRCRAFT TYPES

1. Fighters

Boeing F-15I Eagle | *Ra'am (thunder)*

A derivative of the F-15E Strike Eagle, the F-15I entered service with the IDF/AF on 19 January 1998. In line with an order confirmed in 1994, a total of 25 jets (an initial 21, followed by four more to provide a full squadron complement) had been delivered by 1999. The F-15I is qualified to launch almost every air-to-air and air-to-ground weapon in the IASF arsenal. The latest addition to the avionics and communication suite is the satellite communication (SATCOM) capability, which provides the aircraft with an efficient communication channel to IASF HQ across very long distances. The F-15I's normal mode of operation in penetration missions is to fly very fast and very low, to avoid enemy radar detection. This is achieved using the LANTIRN (Low-Altitude Navigation and Targeting Infrared for Night) system, allowing low-altitude flying under all weather/visibility conditions to attack ground targets with a variety of precision-guided weapons. The IASF operates the F-15I within a single unit, No. 69 'Hammers' Squadron, stationed at Hatzerim.

F-15I serial number 244. This July 2011 photograph shows an unusual air superiority configuration, with a mix of Python 5 and AIM-120 AMRAAMs under the wings and AIM-7 Sparrow AAMs below the conformal fuel tanks. (Ofer Zidon)

Modern Israeli Air Power

A photo from the F-15I display routine at Hatzerim, revealing details of the upper-fuselage paint scheme. (Ofer Zidon)

General Dynamics (Lockheed Martin) F-16A/B Fighting Falcon | *Netz (Sparrowhawk)*

The first batch of F-16A/B fighters, known as the Netz 1 in Israeli use, entered service on 2 July 1980. The second batch, named Netz 2, followed on 1 August 1994. The Netz 1 airframes were newly built, while the Netz 2 aircraft were surplus US Air Force and Air National Guard airframes.

In total Israel received 67 F-16As and eight F-16Bs. These were followed by additional deliveries taken from USAF stocks, and comprising 36 F-16As and 14 F-16Bs.

The 'glory days' of the Netz were in the early 1980s when the type participated in the 1981 attack on Iraq's nuclear reactor and posted claims for dozens of Syrian MiGs shot down over the Bekaa Valley in the First Lebanon War in 1982.

With the arrival of the advanced F-16C/D the Netz switched its role. Both single-seat and two-seat versions now fly together and share responsibility for postgraduate training of future IASF fighter pilots.

Chapter 4

F-16A tail number 239 from No. 140 'Golden Eagle' Squadron takes off during a deployment to Ovda. (Ofer Zidon)

F-16B serial number 017 from No. 116 'Defenders of the South' Squadron. Reflecting its Advanced OTU role, the use of two-seat aircraft is very common within the squadron. (Ofer Zidon)

Modern Israeli Air Power

The Netz community includes two squadrons. No. 116 'Defenders of the South' is based at Nevatim and is responsible for the Operational Training Unit (OTU) and Advanced OTU courses for the IASF, while another squadron – No. 115 'Flying Dragon' Squadron or the 'Red Squadron' – is charged with advanced training and the adversary role with the Advanced Training Center at Ovda. A third unit, Nevatim-based No. 140 'Golden Eagle' Squadron, handed over its airframes to the co-located No. 116 Squadron in August 2013. No. 140 'Golden Eagle' Squadron will subsequently re-form as the launch F-35 operator. The Netz still takes part in the IASF fighter squadrons' alert rotation, assuming responsibility for the air superiority mission and for dropping illumination bombs.

General Dynamics (Lockheed Martin) F-16C/D Fighting Falcon | *Barak (lightning)*

Five squadrons of the IASF are equipped with the F-16C/D. Three of these squadrons operate the older, 1987-vintage F-16 Block 30 (Barak 1), of which deliveries amounted to 51 F-16Cs and 24 two-seat F-16Ds. Two squadrons operate the newer F-16 Block 40 (Barak 2) delivered from 1991, in the form of 30 F-16Cs and 30 F-16Ds. The force is divided into homogenous single-seat and two-seat squadrons. Ramat David's No. 109 'Valley' Squadron is a two-seat squadron, while No. 110 'Knights of the North' and No. 117 'First Jet' Squadrons are single-seat outfits. Hatzor's squadrons followed Ramat David in 2004, with the exchange of single- and two-seat aircraft among No. 101 'First

F-16C serial number 364 from No. 110 'Knights of the North' Squadron takes off from Ramat David. The fighter wears a kill marking on its nose, applied after it shot down a Hezbollah UAV during the 2006 Second Lebanon War.
(Ofer Zidon)

F-16D serial number 036 from No. 109 'Valley' Squadron lands after a QRA mission. The aircraft is armed with Rafael Python 4/5 and AIM-9 AAMs. The enlarged spine contains electronic equipment of Israeli design and manufacture.
(Ra'anan Weiss)

Fighter' and No. 105 'Scorpion' Squadrons, making the former a single-seat squadron and the latter a two-seat squadron.

The difference in hardware between single- and two-seat squadrons also dictates a difference in squadron missions. The single-seat squadrons specialise in air superiority missions and the delivery of autonomous precision-guided munitions such as the JDAM and Rafael Spice, while the two-seat squadrons specialise in the employment of TV- and laser-guided precision weapons such as the GBU-10, GBU-15 and Popeye.

The Barak fleet recently completed an avionics and communications upgrade programme known as Barak 2020, aimed at bringing these components up to the same standard as the F-16I. Together with the newer F-16I, the F-16C/D fleet is likely to play an important part in any future conflict, regardless of the opponent.

Lockheed Martin F-16I Fighting Falcon | *Sufa (storm)*

Between February 2004 and February 2009, a total of 102 F-16I fighters were delivered to Israel. They equipped four squadrons: No. 119 'Bat', No. 201 'The One' and No. 253 'Negev' Squadrons operating from Ramon, and No. 107 'Orange Tail' Squadron operating from Hatzerim alongside the F-15Is of No. 69 'Hammers' Squadron.

The F-16I is a modified two-seat F-16D Block 52 with many Israeli-developed avionic systems installed, including navigation and targeting systems, mission control and communication systems including SATCOM, electronic warfare and self-protection systems, the Elbit DASH helmet-mounted display and target acquisition system, and others. The F-16I uses the same Pratt & Whitney F100-PW-229 engines as the F-15I and with its upper-fuselage conformal fuel tanks has almost the same range.

The F-16I and F-15I complement each other in a long-range operational scenario. Since the F-16I is a two-seat aircraft, the weapons systems operator (still called a navigator in IASF service) is able to focus on navigation, target acquisition and weapons delivery. In terms of long-range missions, the main weakness of the F-16I is its single engine and inferior weapons load, the latter factor offset by the large number of aircraft in the IASF inventory.

Modern Israeli Air Power

F-16I serial number 457 from No. 119 'Bat' Squadron takes off from Ramon. Note the grey SATCOM cover on the rear of the aircraft's spine.
(Ofer Zidon)

F-16I serial number 493 from the 'Bat' Squadron lands at Ramon.
(Ofer Zidon)

44

Chapter 4

McDonnell Douglas (Boeing) F-15A/B/C/D Eagle | *Baz (buzzard)*

The IASF's fleet of F-15A/B/C/D fighters populates two squadrons, No. 106 'Spearhead' and No. 133 'Twin Tail', both based at Tel Nof. The Baz has flown with the Israeli Air Force since 1976 and despite its four decades of service it will not be replaced in the foreseeable future.

Israel placed its first order for the F-15A/B in 1975, and these aircraft were delivered under the Peace Fox I and II contracts signed in 1976. Originally, it seems the IDF/AF requested 50 F-15s, and Washington offered 48. Eventually, funds were limited, and therefore orders and deliveries proceeded at a somewhat irregular rate. Peace Fox I comprised four F-15A AFDT&E (Air Force Developmental Test and Evaluation) airframes, 19 production F-15As and two F-15Bs for a total of 25 aircraft. Of these early F-15As, at least four have since been withdrawn from service and act as gate-guards. Peace Fox II, delivered by April 1982, encompassed nine F-15Cs and six F-15Ds, for a total of 15 jets. An additional nine F-15Cs and two F-15Ds (11 aircraft in all) followed under Peace Fox III that was delivered from April 1985. Peace Fox IV provided 25 ex-USAF Eagles, comprising 19 F-15As and six F-15Bs that had not undergone the Multi-stage Improvement Program (MSIP) upgrade; these aircraft were delivered from October 1991. In the event, a number of the Peace Fox IV aircraft never entered service, but were used as a source of spares, and it appears that of the 17 aircraft that did enter service all but one had been retired by the IDF/AF by 2001. Peace Fox V saw five F-15Ds ordered in 1988 and delivered in May 1992 (in fact, the airframes involved are understood to be F-15Es). Finally, attrition replacements are understood to include a single ex-USAF F-15A (or perhaps an F-15C) delivered together with the Peace Fox III aircraft.

The above deliveries provide a total of 43 F-15As, eight F-15Bs, 18 F-15Cs and 13 F-15Ds for a grand total of 82 aircraft. Of these, it is likely that as of 2013 at the very most 60, but more likely around 50 remain in service with the two units.

F-15A serial number 654 from No. 133 'Twin Tail' Squadron taxis back to its HAS at Ovda during a squadron deployment to the base.
(Ofer Zidon)

Modern Israeli Air Power

F-15D serial number 280 from the 'Twin Tail' Squadron. The fighter's individual name, written on the port side of the nose, is *Yad Hanefetz* (shatterhand). The F-15 is adorned with a roundel commemorating its participation in Operation Wooden Leg – the 1985 attack on the PLO HQ in Tunis, which remains the longest-range attack mission performed by the Israeli Air Force.
(Ofer Zidon)

F-15C serial number 802 from No. 106 'Spearhead' Squadron. The aircraft wears the individual name *Panter* (panther) and carries four kill markings indicating Syrian MiGs claimed shot down in the 1982 First Lebanon War. The aircraft is on Quick Reaction Alert (QRA) and is armed with Python 3, Python 5, and AIM-7 missiles, as well as a single ECM pod.
(Ofer Zidon)

F-15B serial number 109 from the 'Spearhead' Squadron lands at Tel Nof.
(Ofer Zidon)

The lack of successor to the Eagle drove Israel to initiate the first F-15 Baz improvement programme – Baz Meshupar (Improved Baz) – in the first decade of the 21st century. The programme has not only enabled the IASF to keep the Baz airframes at the front line of its interceptor force, but also provided them with new air-to-ground capabilities including the delivery of precision-guided weapons such as Joint Direct Attack Munition (JDAM) and Rafael Popeye missile.

Even though the F-15 is still the only IASF interceptor that can engage in combat with high-altitude, low wing-loading threats such as the MiG-25 and Sukhoi Su-27 family, its air-to-ground capability was successfully exploited in the 2006 Second Lebanon War and the 2009 Cast Lead operation in Gaza.

The F-15 Baz fleet is now entering a new improvement programme, known as Baz La'ad (Baz Forever), aiming at extending the life of the interceptor for at least another 10 years. The improvements include the replacement of tens of kilometres of wiring, replacement of the original control surfaces with lighter and stronger gridlock-type surfaces and addressing problems related to the control surface actuators.

Fighter fleet renewal

In September 2008 the IASF selected the Lockheed Martin F-35 Lightning II Joint Strike Fighter as its next fighter aircraft. Initially, the Lightning II is expected to replace the veteran F-16A/B fleet.

The F-35 is not the ideal choice as the IASF's next-generation fighter asset, as a result of its rigidly dictated avionics suite and the lack of a two-seat variant. Indeed, Israel informed the US Department of Defense that a prerequisite to its purchase was the replacement of 50 per cent of F-35 systems with Israeli-manufactured technology. Israel had previously indicated its intention to integrate locally developed radar and other electronic warfare systems in the F-35, as well as to provide the aircraft's primary flight computers.

Israel's dependence on US funding left it little choice but to issue a 2009 request for purchase of an initial batch of 20 F-35I conventional take-off and landing (CTOL) aircraft, with an option to purchase an additional 50 aircraft at a later date – perhaps to include the short take-off and vertical landing (STOVL) F-35B version. In October 2010 the Israeli government formally approved the planned USD 2.75-billion acquisition for 20 aircraft and signed a Foreign Military Sales (FMS) agreement with the US.

IAI has also signed a contract with Lockheed Martin to produce wings for the F-35, using a new production line at its Lahav facility.

The IASF has already decided on the Hebrew name of the new fighter – Adir (mighty) – and the squadron's location will be Nevatim, the current home of the bulk of the F-16A/B force.

With the F-35 programme subject to delays, in 2011 it was reported that Israel was considering stopgap solutions to maintain its fighter fleet at the desired level until the first Lightning IIs are delivered. Options under consideration reportedly include the purchase of used F-15s from the USAF and further upgrades for its existing fleet of F-15s and F-16s. Delays in the Joint Strike Fighter programme could postpone the delivery of Israel's first F-35s until 2018.

2. Transports

The IASF transport aircraft force is divided into two wings. The Heavy Transport Wing is located at Nevatim in the southern Negev Desert and contains two C-130 squadrons, No. 103 'Flying Elephant' and No. 131 'Yellow Bird' Squadrons; one Boeing 707 squadron, No. 120 'Desert Giants', which also flies the IAI Seascan; and one Gulfstream special missions squadron, No. 122 'Nachshon' Squadron. In order to host the heavy transport aircraft, Nevatim was upgraded over the last decade, to include modern facilities including a new control tower and runway – the longest in the Middle East – aprons, workshops, depots and others.

The Light Transport Wing is based at Sde Dov on the northern outskirts of Tel Aviv. Beginning in the first half of the 1990s, the Light Transport Wing underwent a complete process of re-equipment. The Light Transport Wing operates two squadrons, No. 100 'Flying Camel' and No. 135 'Kings of the Air' Squadrons, which fly the Beechcraft King Air 200, Beechcraft RC-12 and the Beechcraft Model 36 Bonanza. Sde Dov has also seen recent infrastructure renewal, including a new control tower, new hangars and new aprons.

With the 707 tanker fleet due for replacement, Israel currently has two options. The favoured platform is reportedly the Boeing 767, converted to refuelling configuration by IAI's Bedek division, and similar to the aircraft already delivered by IAI to the Colombian Air Force, and as selected by the Brazilian Air Force. Alternatively, the US is offering surplus USAF KC-135. The latter option would likely be dependent upon Israel placing an order for the V-22 Osprey tiltrotor, or upon a new round of peace talks with the Palestinian Authority.

Beechcraft Model 36 Bonanza | *Hofit (stint)*

The Beechcraft Model 36 Bonanza entered service on 5 December 2004 and replaced the short-lived SOCATA TB-20 Trinidad (known locally as Pashosh – warbler) in the light transport and air taxi missions. These aircraft all receive maintenance and technical support with the Light Transport Wing at Sde Dov, although examples are stationed at various airfields to serve the liaison needs of the local base.

Bonanza serial numbers 346 and 304 from No. 135 'Kings of the Air' Squadron parked at Nevatim.
(Ofer Zidon)

Chapter 4

Beechcraft Model 200 King Air | *Tzufit (sunbird)*

Due to technological advances in visual intelligence (VISINT) equipment and the development of new and heavy electro-optical payloads, the Israeli Air Force was forced to replace its old light transport and VISINT platforms – the Dornier Do 28, or Agur (crane) and the Beechcraft Model 80 Queen Air, or Zamir (nightingale). The replacement aircraft was required to carry heavier payloads and fly at higher altitudes. The Israeli Air Force selected the Beechcraft Model 200 King Air, mainly because it had already used the military version – the Beechcraft RC-12 – since 1984.

The new aircraft was introduced to IDF/AF operations in 1991. In total, Israel received 22-23 Model 200CT/T versions. The twin-engine turboprop assumed the roles of the C-47, Queen Air, IAI Arava and the Do 27 Agur, including cadet training, intelligence gathering, observation, and command and control missions. The unification of so many roles and aircraft types into a single airframe reduced demands on maintenance and manpower and simplified the infrastructure required to support the wing. For many years the Tzufit fleet was the busiest within the air force, flying the highest number of hours per year.

King Air serial number 709 is on strength with the appropriately named No. 135 'Kings of the Air' Squadron.
(Ofer Zidon)

King Air serial number 714 from the 'Kings of the Air' Squadron. Note that the aircraft retains its civilian colours and was not painted in the standard overall grey scheme.
(Ofer Zidon)

49

Modern Israeli Air Power

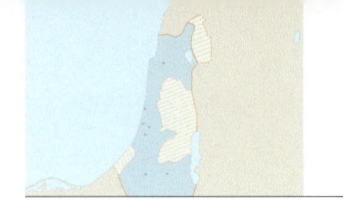

Boeing 707 | *Re'em (oryx)*

Beginning in 1978 the IDF/AF replaced its ageing Boeing 707-100 fleet with newer 707-300 airframes. It is understood that Israel received six 707s configured as tankers, plus six electronic intelligence versions. The final few ELINT 707s were all withdrawn following the arrival of the Gulfstream in 2006. At present the main role of the 707 is aerial refuelling, and a conversion process of 707 transport airframes to the tanker configuration is due to end in the near future. The current fleet therefore comprises eight tankers, the last of which entered the conversion phase in late 2008, with IAI carrying out a USD 23-million conversion. The IAI conversion process includes installation of new fuel tanks, fuel pumps and a locally developed electro-optical assisted boom system to allow accurate refuelling of the F-16 and F-15 fighters.

The entire 707 tanker fleet is also undergoing a service life extension programme that includes the addition of new communications and avionics equipment. The aircraft are therefore receiving new 'glass' cockpits and upgraded communications equipment from IAI, with the first upgraded aircraft having been re-delivered in November 2009. Included in the new communications suite is SATCOM equipment, evidenced by a prominent black radome on the upper fuselage.

The 707 plays an important part in any IASF long-range power-projection scenario and therefore all IASF fighter squadrons devote many training sorties to practising aerial refuelling with the 707 fleet. The aircraft also accompanies every IASF deployment abroad.

Boeing 707 serial number 264 from the 'Desert Giants' Squadron painted in the dark grey scheme. The latest colours are already fading under the desert sun.
(Ofer Zidon)

Lockheed C-130E/H and Lockheed Martin C-130J Hercules |
Karnaf (rhinoceros) and Shimshon (Samson)

The C-130 assumes the heavy transport role in IASF operations. Total deliveries comprised 10 C-130Hs (including one KC-130 tanker two EC-130 electronic warfare platforms), two KC-130Hs, and 12 former USAF C-130Es. The ageing fleet of 1970s-era C-130E/Hs will soon receive a long-awaited reinforcement in the shape of newly manufactured C-130Js. A March 2010 visit by Defence Minister Ehud Barak to the US saw approval provided for the purchase of up to nine C-130Js, in addition to the new F-35 fighters. In April 2010 Israel ordered a single 'stretched' C-130J-30 via Foreign Military Sales channels, followed by a second ordered in April 2011, a third in February 2012 and a fourth in July 2013. Meanwhile, formal delivery of the first C-130J-30 for the IASF was marked at Lockheed Martin's Marietta facility on 26 June 2013. The aircraft was expected to arrive in Israel in spring 2014.

The C-130 fleet was traditionally operated by two squadrons: No. 103 'Flying Elephant' and No. 131 'Yellow Bird' Squadrons. The two squadrons previously shared the airframes, so some aircraft tails are adorned with both squadron badges, the 'Yellow Bird' on the starboard side and the 'Flying Elephant' to port. With the arrival of the C-130J in 2014, the Hercules fleet will be separated, the 'Flying Elephant' Squadron operating the new C-130J while the 'Yellow Bird' Squadron continues to fly the older C-130E/H. The veteran C-130E is now operated in dwindling numbers, and according to technical availability. As of August 2013, all C-130E/H aircraft had been amalgamated into the 'Yellow Bird' Squadron, allowing the 'Flying Elephant' crews to prepare for the arrival of the C-130J.

A C-130H demonstrates its short-field capability with the aid of a rocket-assisted take-off (RATO). This allows the big transport to operate from combat-zone airstrips. Note the 'Yellow Bird' Squadron insignia on the starboard side of the tail. (Ofer Zidon)

Modern Israeli Air Power

3. Attack helicopters

The Israeli attack helicopter force was built up according to lessons learned in the October 1973 War and the requirement to blunt assaults by massed enemy armour. Over the years the Israeli Air Force attack helicopter force has evolved from dedicated tank-hunters to gunships capable of 'surgical strike' to address new threats and attack guerrilla forces positioned within civil populations.

The attack helicopter community consists of two front-line squadrons: No. 190 'Magic Touch' Squadron flying the AH-64A and No. 113 'Hornet' Squadron flying the AH-64D, both based at Ramon. These squadrons are joined by the veteran AH-1 fleet. In its prime the Tzefa fleet consisted of two front-line squadrons, a rotary-wing element within No. 115 'Flying Dragon' Squadron which serves as the adversary unit at Ovda, and an attack helicopter squadron for advanced training within the Air Force Academy. In recent years the AH-1 fleet has been reduced, with emphasis on training. The final front-line AH-1 operator, No. 160 'First Attack Helicopter' Squadron, was disbanded at Palmachim in August 2013.

The AH-64 fleet is slowly undergoing a transformation process aimed at converting the veteran AH-64A airframes to the later AH-64D standard. AH-64A airframes are shipped to Boeing's installations in the US where they are rebuilt to AH-64D standard.

Bell AH-1E/F Cobra | *Tzefa (viper)*

The AH-1 entered IDF/AF service in 1975. At first, Israel received six ex-US Marine Corps AH-1G helicopter gunships and used them to develop combat doctrine for the new platform. No. 160 'First Attack Helicopter' Squadron was established in 1977,

AH-1F serial number 452 from No. 160 'First Attack Helicopter' Squadron is armed with the Elbit Systems Machtselet missile. This long-range electro-optical missile is part of the Rafael Spike family. (Ofer Zidon)

equipped with a dozen newly built AH-1s and a number of Defender helicopters. The newly developed doctrine was put to a successful test in the 1982 First Lebanon War, fighting Syrian armoured divisions in Lebanon. Large-scale Cobra deliveries involved over 30 TOW missile-equipped AH-1Fs between 1983 and 1990, which led to the initiation of a second squadron – No. 161 'Southern Cobra' Squadron – in 1985. Over a dozen ex-US Army AH-1Es arrived in Israel during 1996 and were allocated to the Air Force Academy, where they continue to serve in the role of training new attack helicopter pilots. With the increase in number of AH-64 attack helicopters in service, the two AH-1 squadrons were amalgamated into No. 160 'First Attack Helicopter' Squadron in 2005.

Although inferior to the AH-64A and AH-64D in terms of engine power and the avionics suite, the AH-1 still possessed capabilities that remain unavailable to the AH-64 family, including the Rafael Machtselet (mattress) weapon system. However, as of August 2013, the AH-1 is restricted to second-line duties, with No. 115 'Flying Dragon' Squadron at Ovda, and in an attack helicopter squadron within the Air Force Academy.

McDonnell Douglas AH-64A and Boeing AH-64D-I Apache |
Peten (python) and Saraf (serpent)

The AH-64D-I is the latest addition to the attack helicopter force. Many Israeli-built systems were integrated into the helicopter, including an Elta communication suit, Elbit mission management system, Rafael Combat Net system and Elisra self-protection suite. The AH-64D-I entered service on 4 April 2005 and the fleet is being doubled with the upgrade of older AH-64A Peten airframes to AH-64D-I Saraf standard.

The first procurement contract for 18 AH-64As was signed in May 1990, with the first shipment of surplus US Army airframes arriving only four months later, on 11 Sep-

AH-64A serial number 842 from No. 190 'Magic Touch' Squadron. Note the fairing for the SATCOM antenna above the port stub wing.
(Ofer Zidon)

AH-64D serial number 736 from No. 113 'Hornet' Squadron flies over the desert in southern Israel. Note the asymmetrical configuration of external fuel tanks and the fact that the typical Longbow radar dome is missing from the top of the rotor.
(Ofer Zidon)

tember 1990. The last of these 18 AH-64As arrived in May 1991, and in September 1992 the US Department of Defense announced the delivery of another 24 AH-64As to Israel. The surplus Apaches were delivered by air on 12 September 1993. The next step in the evolution of the fleet took place in December 2000 with the procurement contract for nine AH-64Ds, eight of them newly built and one converted from an existing AH-64A, shipped to the US. With the success of the conversion process proven, two additional conversion contracts were signed, one for the conversion of three helicopters signed in April 2003, and a second signed in August 2004 for the conversion of another six helicopters. The last step in the evolution of the AH-64 fleet took place in early 2010 with the signing of another conversion contract for three helicopters, aimed at replacing three helicopters lost during the 2006 Second Lebanon War. AH-64A serial numbers 822, 929 and 953 were shipped to Boeing and the converted AH-64Ds (serial numbers 787, 113 and 789) arrived in Israel in October 2012. In total, therefore, 13 AH-64As have been converted to AH-64D standard.

Chapter 4

4. Assault helicopters

The IASF operates two types of assault helicopters, the Sikorsky UH-60 for medium lift and the Sikorsky CH-53 for heavy lift. The main role for both types is troop transport. They also conduct search and rescue missions, which usually involve Unit 669, which provides the helicopters with soldiers and medical teams according to the particular mission. Since both platforms can carry out search and rescue missions, the decision as to which platform to use is based on the distance and severity of the event and the number of casualties. Some of the IASF assault helicopters carry special markings below the starboard cockpit window to indicate the number of times they have crossed Israel's borders on operational sorties.

Israel has long been viewed as a likely customer for the CH-53K, the latest iteration of the CH-53 family, although the CH-47 is an alternative option. Meanwhile, the V-22 tiltrotor has been highlighted under plans to expand capabilities in the field of long-range commando operations, and the insertion and extraction of special forces.

Sikorsky S-65C and CH-53 Sea Stallion | *Yasur (petrel)*

The heavy-lift CH-53 entered IDF/AF service back in 1969, at the height of the War of Attrition. The fleet includes the survivors from around 38 original S-65C-2/3 airframes, two former Austrian Air Force S-65Ös, and some 24-26 additional CH-53A airframes delivered to Israel from surplus US Marine Corps stocks in two batches – one in 1974 and the second in 1991. With no real alternative, the ageing and very busy Yasur fleet was upgraded to Yasur 2000 standard during the 1990s and to Yasur 2025 standard in the 2000s. The first programme involved bringing CH-53A airframes up to the CH-53D

CH-53 Yasur 2000 serial number 036 from No. 114 'Night Leaders' Squadron. The squadron insignia is painted on the tail and mission markings have been applied under the cockpit window, each mark representing an operational sortie behind enemy lines.
(Ofer Zidon)

CH-53A Yasur 2025 serial number 912 from No. 114 'Night Leaders' Squadron. (Ofer Zidon)

standard, including the addition of a refuelling probe. Both improvement programmes addressed construction issues, installation of modernised avionics and standardisation of the interior and cockpit. Part of the standardisation process was the addition of external fuel tanks to the CH-53A, with external support rods for the tanks. The Yasur 2025 programme added a new self-protection system, new electronic warfare system, new communication system with a datalink, and a new terrain avoidance system. As of summer 2013, both Yasur 2000 and 2025 versions remained in service, although eventually all will be brought up to the later standard. Two IASF squadrons are equipped with CH-53s: No. 114 'Night Leaders' Squadron and No. 118 'Night Predators' Squadron, both operating from Tel Nof.

Sikorsky S-70A/UH-60 Black Hawk | *Yanshuf (owl)*

The UH-60 entered IDF/AF service in 1994. The Black Hawk is the IASF's medium assault helicopter, capable of transporting up to 10 soldiers or internal and external (underslung) loads. As well as troop-transport missions, the UH-60 is used for medical evacuation, for which is will normally embark soldiers and medical staff from Unit 669. Other roles include supply of front-line units, and search and rescue.

The first to arrive were the 10 examples of ex-US Army UH-60A variant, or Yanshuf 1, while the ex-US Army UH-60L Yanshuf 2 arrived in 1998, and a final batch of new-build

Chapter 4

UH-60L serial number 856 wears US Army Olive Drab colours. Operated by No. 124 'Rolling Sword' Squadron, the ex-US Army Black Hawk originally received the serial number 956. It became 856 after modification in Israel, but retained its original colours. (Ofer Zidon)

UH-60L Yanshuf 2 serial number 830 from No. 124 'Rolling Sword' Squadron. (Ofer Zidon)

UH-60L Yanshuf 3 serial number 517 from the 'Desert Owls' Squadron. Upgrades have seen the Israeli UH-60 fleet has been equipped with a comprehensive self-protection suite. (Ofer Zidon)

UH-60L Yanshuf 3s (manufacturer's designation S-70A-50/55) were purchased and delivered in 2002. In total, Israel has received 35 UH-60Ls in the two aforementioned batches. Ongoing modifications have kept the Yanshuf fleet up to date, and include a mission debrief system, aircraft health management system, self-protection system and new hoist. Some of the UH-60 airframes have been fitted with external fuel tanks under the mid-section stub wings and an aerial refuelling probe under the port side of the cockpit. The UH-60 fleet is split between two squadrons: No. 123 'Desert Owls' Squadron at Hatzerim and No. 124 'Rolling Sword' Squadron at Palmachim.

During 2008 an armed version of the UH-60 was tested by the IASF. The 2006 conflict in Lebanon emphasised the robustness of the airframe and showed that an armed Black Hawk might be useful in both the traditional troop transport role and the attack role. The wide fuselage could house weapons and related avionics systems, while the generous fuel capacity and aerial refuelling capability make it suitable for attack missions deep into enemy territory. The IASF loaned one of its UH-60Ls to the Armed Black Hawk project and Elbit Systems was appointed to lead the conversions required to accommodate the new systems. The helicopter was armed with an air-to-ground missile system, most likely a Rafael Spike family member or new derivative, as well as a cannon under the nose. In the event, no purchase was made.

Chapter 4

5. Special missions

Air Tractor AT-802F | *Matar (rain)*

A forest fire on Mount Carmel from 3 to 5 December 2010 claimed the lives of 44 people. The fire saw an armada of firefighting aircraft gather in Israel, including military and civilian assets from Cyprus, France, Greece, Russia, Turkey, the UK and the US. The decision was taken to procure a firefighting aircraft, which would be operated by a dedicated new unit, No. 249 'Elad' Squadron.

Several options were considered, including re-qualifying the helicopter force with water-bombing capabilities, the purchase of Modular Airborne Fire Fighting System (MAFFS) kits for use with the C-130 fleet, or the purchase of specialised firefighting aircraft. The decision made was to buy firefighting aircraft, with the purchase of the Bombardier CL-415 as the preferred long-term solution.

In the meantime, a stopgap solution was introduced, in a form of a contract issued to Elbit Systems under which the company would supply flight hours for Air Tractor AT-802F firefighting aircraft. In turn, Elbit chose Chim-Nir, a supplier of aerial agricultural applications, specialised in crop-dusting missions, fly the Air Tractors.

In order to fulfil the contract, Elbit purchased eight Air Tractors. Five are known to have arrived from Aviavilsa in Valencia, Spain, including two AT-802F Fireboss equipped with floats to allow scooping of seawater into the aircraft's three-ton water tank, and at least two from Air Tractor in the US.

Initial operational use of the AT-802F made it clear that aircraft designed to extract water from the sea were useless because of Israel's sea conditions, which were not calm enough for the scooping operation. As a result, the two floatplanes were converted to wheeled undercarriage.

The AT-802Fs are primarily based at Meggido airfield, from where they are operated by Chim-Nir Aviation on behalf of the ISAF.
(Erez M. Sirotkin)

Beechcraft RC-12 | *Kukiya (cuckoo)*

The first Beechcraft RC-12 aircraft were delivered to the IDF/AF in 1984. They are operated by No. 135 'Kings of the Air' Squadron from Sde Dov and specialises in SIGINT missions, including search and rescue of downed pilots over enemy territory. The high volume of missions required from the RC-12 fleet forced the IDF/AF to buy another batch of aircraft in the early 1990s. The existence of these aircraft was unveiled only as late as June 1996 and some of its missions remain classified. In terms of deliveries, reports indicate that Israel received six RU-21Ds, three RU-21As and four RC-12Ds.

Beechcraft RC-12 serial number 987 from No. 135 'Kings of the Air' Squadron. The electronic warfare version of the King Air is identified by its vast array of antennas.
(Ofer Zidon)

Eurocopter AS.565MA Panther | *Atalef (bat)*

The AS.565MA entered IDF/AF service in 1996 as a dedicated maritime helicopter in support of the Israeli Navy's Sa'ar 5 (tempest) class of missile corvettes. Although the fleet of five AS.565MAs is operated and maintained by the IASF within No. 193 'Defenders of the West' Squadron based at Ramat David, its operations are financed by the Israeli Navy and a naval officer is an integral part of the helicopter's crew, coordinating operations between the IASF flight crew and the Navy crews at sea. One of the squadron's helicopters was damaged during the Second Lebanon War in 2006 when a Sa'ar 5 corvette was hit by Chinese-made C-802 coastal defence missile fired by Hezbollah.

AS.565MA serial number 882 from the 'Defenders of the West' Squadron on approach to Ramat David.
(Ofer Zidon)

Chapter 4

Gulfstream V and 550 | *Nachshon Shavit (pioneer – comet) and Nachshon Eitam (pioneer – fish eagle)*

The IASF's current special mission aircraft are the Gulfstream V (the successor to the Grumman OV-1D Mohawk) and Gulfstream 550 (the successor to the Grumman E-2C Hawkeye). The new aircraft also inherited some of the electronic warfare missions previously flown by the Boeing 707. Both Gulfstream types carry the family name of Nachshon. The Gulfstream V, or Nachshon Shavit (pioneer – comet), is fitted with ground-scanning radar and is configured for the signals intelligence (SIGINT) mission and entered service in June 2005. The Gulfstream 550 is fitted with phased-array radars on both sides of the forward fuselage and is configured for the airborne early warning (AEW) mission. It is known as the Nachshon Eitam (pioneer – fish eagle) and entered service a year after the Shavit, in September 2006. Both types are fitted with mission equipment made in Israel by IAI's Elta division. The exact number of Nachshons received by the Israeli Air Force remains a closely guarded secret. However, it is understood that the fleet comprises three Nachshon Shavits, and three or four Nachshon Eitams.

Both Gulfstream Nachshon types from No. 122 'Nachshon' Squadron in formation. The upper aircraft is the Gulfstream 550 Nachshon Eitam airborne early warning aircraft and the lower is the Nachshon Shavit SIGINT aircraft, based on the Gulfstream V airframe. (Ofer Zidon)

61

Modern Israeli Air Power

IAI 1124N Seascan | *Shachaf (seagull)*

The IAI 1124N Seascan is a maritime patrol and observation platform based on the IAI 1124 Westwind I business jet. Operations and coordination of the Seascan are conducted in the same way as the AS.565MA. Only three examples of the Seascan remain in use and these will likely be retired in the foreseeable future to be replaced by a maritime version of the IAI Heron UAV.

Seascan serial number 927 from No. 120 'Desert Giants' Squadron is seen conducting a maritime patrol sortie. (Ofer Zidon)

6. UAVs

Israel's unmanned aerial vehicles (UAV) were originally developed as a small and efficient platform for real-time intelligence-gathering missions over enemy air defences, required because of the threat posed by mobile air defence systems like the Soviet SA-6. Since its introduction to the IDF/AF in 1979, the capabilities of the UAV fleet have evolved tremendously, with the development of new and advanced payloads allowing the UAV to develop from being the response to a specific threat, to taking over a wide range of roles, from simple VISINT missions to communication relay and maritime patrol. The evolution in UAV force structure and missions is manifested by the development of the fleet from a single type in 1979 to multiple types and multiple squadrons today. Currently, the IASF operates UAV systems in intelligence-gathering missions and attack missions. Future development concepts cover a rotary-wing UAV, maritime missions UAV, casualty evacuation UAV and an aerial tanker UAV.

Elbit Hermes | *Zik (spark)*

IAI products dominated the Israeli UAV fleet from 1979 until 1999, with the Scout and Searcher types. In 1999 Elbit Systems introduced the Hermes 450 that entered IDF/AF service in the same year. Follow-on orders were subsequently placed in 2003 and 2007. Ordered in 2010, the Hermes 900 is a development of the Hermes 450. The Hermes 900 flies at an altitude of over 30,000ft (9,144m), has an large payload capacity and improved adverse-weather capabilities.

Chapter 4

Hermes 450 serial number 376 from No. 166 'Hermes' Squadron on display at Hatzor. Note the tail art and Hermes emblem on the nose.
(Ofer Zidon)

IAI Heron | *Shoval (trail)*

Earlier Israeli Air Force UAV systems were relatively short-ranged, with a maximum operational endurance of between 12 and 24 hours and the ability to lift a single payload. In 2007 the picture changed dramatically with the introduction of the one-ton class IAI Heron as a replacement for the ageing Searcher. The Heron can carry two payloads to an altitude of 30,000ft (9,144m) and has an endurance of 36 hours or more.

Heron serial number 255 from No. 200 'First UAV' Squadron in its shelter at Palmachim.
(Ofer Zidon)

63

Modern Israeli Air Power

IAI Heron TP | *Eitan (firm)*

A major advance occurred in 2010 with the introduction of the IAI Heron TP. The Heron TP is a four-ton class UAV powered by a turboprop engine. Its dimensions, range and payload size allowed the IASF to undertake a new spectrum of long-range missions, a fact that called for the creation of a new squadron, No. 210 'Eitan' Squadron. The 'Eitan' Squadron left the historical UAV base at Palmachim and operates from the main air base in central Israel, Tel Nof, which already hosts F-15 Baz and CH-53 squadrons. The new base offers a longer runway and the more extensive infrastructure required for the Eitan, with its 26m (85.3ft) wingspan.

Heron TP serial number 214 from No. 210 'Eitan' Squadron on display at Tel Nof. The soldier standing beside it provides an indication of the size of the UAV.
(Ofer Zidon)

7. Trainers

The IASF training fleet is concentrated at Hatzerim, the home of the Air Force Academy. The Academy includes both fixed- and rotary-wing training syllabuses. Looking to the future, the IASF will replace its veteran A-4 trainers with the new M-346 Master, selected by the IASF in February 2012 in favour of the Korea Aerospace Industries T-50 Golden Eagle. Israel will acquire a total of 30 M-346s under a deal worth around USD 1 billion, in which the trainers will be purchased and maintained by the Thor joint venture, consisting of IAI and Elbit Systems. Thor will then sell 'air time' to the IASF. The IASF will name the M-346 as the Lavi (young lion) – reusing the name of IAI's abortive lightweight fighter. The first M-346s are expected to join the IASF in 2014–15 and will likely be assigned to No. 102 'Flying Tigers' Squadron.

Beechcraft T-6A Texan II | *Efroni (lark)*

The Fouga Magister, and subsequently the upgraded Tzukit (thrush) was used for basic and advanced training of air cadets following its introduction in 1960, with the A-4 taking over the role of advanced trainer from 1972. After the arrival of the Grob 120, the second step in the modernisation of the Academy took place in July 2009, with the arrival of the first four of 20 Beechcraft T-6A Texan II trainers. The new turboprop-

Chapter 4

T-6 serial number 473 from the Air Force Academy's basic training squadron taxis to the end of the runway at Hatzerim. (Ofer Zidon)

powered trainer replaced the Tzukit after 50 years of service. The T-6 was chosen after it won the US Joint Primary Aircraft Training System (JPATS) competition. It was convenient for Israel to purchase a US-made aircraft using Foreign Military Financing (FMF) funding, assuming the US airframe of choice met Israeli needs. The Texan II has many advantages, including an instrument flying capability, an advanced avionics package that is similar to other IASF aircraft, and an ejection seat to increase safety margins.

Bell 206B JetRanger | *Sayfan (avocet)*

Rotary-wing basic training at the Air Force Academy is completed using the Bell 206B JetRanger, between six and 12 of which are in use. The JetRanger entered IDF/AF service in 1971 as a dual-role helicopter, flying both operational and training and air taxi missions. In recent years, following the build-up of the UH-60 force, the JetRanger has become a pure trainer. The change of role led to a change of colour scheme, from the overall brown front-line camouflage to the red and white high-visibility training colours of the Academy.

Bell 206 serial number 123 from the Academy's rotary-wing basic training squadron at Hatzerim. Two unique features of the type in Israel are the extended exhausts and the housing for the manually operated landing lights below the forward fuselage.
(Ofer Zidon)

Grob 120 | *Snunit (swallow)*

The Air Force Academy's syllabus includes a short phase of screening, which serves test to the candidate's aptitude for flying. For many years the screening phase was completed using the veteran Piper Super Cub. The modernisation of the Academy's fleet, which began in 2002, saw Elbit Systems win a contract to sell the Academy flying hours on the Grob 120A-2 for the screening phase, a total of 20 of the new aircraft replacing the 55-year-old Pipers. The Grob 120 features side-by-side seats for the instructor and cadet, a new debriefing system and modern avionics and control panel, to support the progress of the cadets to the advanced phases of the flying course.

Grob 120 serial number 936 4X-DGG flying over Hatzerim. (Ofer Zidon)

McDonnell Douglas A-4 Skyhawk | *Ahit (eagle)*

More than 300 A-4s were purchased by Israel in the late 1960s and early 1970s. The A-4 took over the role of advanced trainer in 1972, four years after it entered service with the IDF/AF and when it still served as a front-line attack platform serving in many operational squadrons. Today, the main role of the A-4 is the training of cadets during the advanced stage of the flight course at the Air Force Academy. However, it also continues to carry out some operational roles, including illumination, and dispensing propaganda leaflets over enemy territory (usually to alert the civilian non-combatant population to evacuate the area).

The A-4 advanced training fleet underwent an upgrade programme beginning in 2005, aimed at extending service life for another 10 years, until 2015. By that time the A-4 will have celebrated almost 50 years in Israeli service and will be replaced by the M-346. The surviving A-4 fleet includes single-seat A-4N and two-seat TA-4H/J serving together with the final operator, No. 102 'Flying Tigers' Squadron at Hatzerim. As of summer 2013, only four examples of the TA-4H version remained active.

Chapter 4

A trio of TA-4Js from No. 102 'Flying Tigers' Squadron at Hatzerim.
(Ofer Zidon)

Bell AH-1E/F Cobra and Sikorsky S-70A/UH-60 Black Hawk

Rotary-wing advanced training at the Air Force Academy is divided into two streams: assault and attack. Advanced assault helicopter training is completed using the UH-60, with aircraft shared with No. 123 'Desert Owls' Squadron based at Hatzerim, while attack helicopter training involves a number of dedicated, non-operational AH-1s. Most of the Academy AH-1 airframes are surplus US Army helicopters that arrived in 1994 and still retain their original dark green colours.

AH-1E serial number 680 from the Air Force Academy. The Academy's AH-1s are ex-US Army machines that still wear the Olive Drab colours of their former operator. Note the brown main and tail rotors and the Academy insignia on the tail.
(Ofer Zidon)

Modern Israeli Air Power

UH-60A serial number 735 from the 'Desert Owls' Squadron hovers as it prepares to land in a field. The aircraft is typical of former US Army Yanshuf 1s now repainted in the later desert colour scheme.
(Ofer Zidon)

Chapter 5

IASF ORDNANCE (CO-WRITTEN BY SARIEL STILLER)

1. Israeli-made weapons

One of the ways in which Israel's air force has maintained its advantage over enemy forces is the quality of its weapon systems. Over the years some of Israel's threats and potential threats have been posed by nations using Western hardware, meaning that Israel's edge has had to be maintained using additions to the airframes themselves. Israel's aerospace industry includes some very large companies led by Israel Aerospace Industries (IAI), Elbit Systems and Rafael Advanced Defense Systems. These companies produce cutting-edge products and weapons systems that are combined within IASF airframes to enhance their functionality and performance. While the air force deploys a wide range of US-made weapons, and various Israeli companies manufacture weapons for export, this chapter focuses exclusively on Israeli-made weapons systems presently in service with the IASF.

1.1. Air-to-air missiles

Rafael Python 3 | Tzafrir (morning wind)
The design and development of the Python 3 missile began in 1978, based on experience gained with the second-generation Shafrir 2 missile. The Python 3 entered service in 1982 and proved itself in air combat during the 1982 First Lebanon War, with claims

An F-15C armed with a Python 3 AAM. The blue body indicates that this is an inert weapon.
(Ofer Zidon)

69

of around 50 kills. The missile's infrared seeker is slaved to the aircraft's radar and initially acquires and tracks the target via this link. The missile is operated in conjunction with Elbit's DASH helmet-mounted display, which allows the Python's seeker to be slaved to where the pilot's head and helmet are directed. The missile is slightly larger than the AIM-9 Sidewinder, thus requiring a specially designed pylon adapter. The missile is carried by all IASF F-15A/B/C/D aircraft.

Rafael Python 4 | Panter (panther)
Development of this short-range all-aspect infrared-homing AAM began in the early 1980s, with entry to operational service in 1992. The missile is reported to have a 60° off-boresight launch capability and can manoeuvre at up to 70g, with the ability to turn 180° after launch and intercept a target behind the launching aircraft. The missile is operated in conjunction with Elbit's DASH helmet-mounted display. The missile is carried by all IASF F-15 and F-16 types.

Rafael Python 5 | Panter 2000 (panther 2000)
A short-range all-aspect infrared-homing AAM, the Python 5 is a modified version of the Python 4 and has been in operational IASF service since the mid-1990s. It retains one of the key characteristics of its predecessors – the seeker is slaved to where the pilot's head and helmet are directed. The missile is carried on all IASF F-15 and F-16 variants. Externally, there is no visible difference between the Python 4 and 5.

A close-up of a Rafael Python 4 or 5 AAM (inboard) under the wing of F-16D serial number 039 from No. 109 'Valley' Squadron. (Ofer Zidon)

All the above missiles are operated in conjunction with Elbit's DASH helmet-mounted display. In close combat prior to helmet-mounted displays, the pilot had to align the aircraft to shoot at a target. DASH enables pilots to aim their weapons simply by looking at the target. DASH measures the pilot's line of sight (LOS) relative to the aircraft and transfers this information to other weapons control systems in the aircraft, slaving the weapons to the target.

Chapter 5

1.2. Air-to-ground weapons

Elbit Opher and Whizzard
Development of this guidance kit began in the early 1980s and it entered Israeli service in 1992. The kit enables general-purpose Mk 82 and Mk 84 bombs to become precision-guided munitions in order to strike pinpoint targets. The kit consists of either a cooled infrared seeker (Opher), or a laser seeker (Whizzard). The seeker is mounted in a stabilised gimballed housing at the end of a probe fitted to the nose of the bomb. Elbit also offers a GPS-guided version of the kit. The bombs can be employed by all IASF strike aircraft.

The Opher laser-guidance kit is seen mated to a 500lb (227kg) Mk 82 bomb.
(Ofer Zidon)

IMI ATAP cluster bombs
The ATAP family includes three bomb sizes: the ATAP-300 that carries 320 bomblets, the ATAP-500 that carries 512 bomblets and the ATAP-1000 the carries 900 bomblets. The cluster bombs can be employed by all IASF strike aircraft.

IMI Delilah | Chanit-Zahav (golden javelin)
Development of the Delilah began in the early 1980s, with service entry following in 1992. The Delilah is an air-launched standoff and cruise missile, with a turbojet engine that allows it to loiter over the battlefield in order to target well-hidden threats. It is also able to attack moving targets, making it ideal for destroying surface-to-air missile systems. The Delilah has a range of 250km (155 miles) and can perform complex manoeuvring using a combination of its engine, aerodynamic structure and electronics. The Delilah incorporates a datalink and can transmit live images back to its operator. It was initially designed for use by Israeli F-4Es and F-16s and is currently in use on IASF F-16C/D/I aircraft.

Modern Israeli Air Power

A Delilah missile seen under the wing of an F-16I from No. 201 'The One' Squadron. (Ofer Zidon)

IAI Griffin

With official service entry in 1991, the Griffin laser-guided bomb system is designed for use with Mk 82, Mk 83 and Mk 84 bombs. The laser-guidance kit is composed of a guidance and control section and an aerodynamic kit. The bomb can be employed by all IASF strike aircraft.

IMI Samson air-launched decoy

The main purpose of the Samson air-launched decoy is to confuse and saturate enemy air defences as part of an overall suppression of enemy air defences (SEAD) strategy, allowing the penetration and attack of highly protected anti-aircraft missile batteries and related systems. The Samson can imitate the flight profile of a fighter aircraft and is also able to dispense chaff over the target area. The decoy is carried by IASF A-4 aircraft.

A Samson decoy on display. (Ofer Zidon)

Chapter 5

Rafael Popeye | Chanit-Na'a (pleasant javelin)
The development of this medium-range, TV-guided air-to-surface missile began in the early 1980s and it entered operational service in 1989. The missile provides the IASF with a standoff capability against high-value targets. The missile has two canard stabilisers, four clipped delta wings and four moving clipped delta fins. Reports suggest the missile has a range in excess of 200km (124 miles) and a 360kg (794lb) warhead. The missile is equipped with a TV datalink antenna that communicates with a datalink pod carried by the launch aircraft. The back-seat weapons systems operator can direct the Popeye to its target, and control over the missile can also be transferred to other platforms after launch. The missile is carried by all IASF F-15 aircraft.

A close-up of a Popeye under the wing of F-15D serial number 797.
(Ofer Zidon)

Rafael Spice | Barad Kaved (heavy hail)
The Spice guidance kit is a derivative of Popeye air-to-surface missile technology. It employs multiple guidance methods and has the ability to be fed, pre-flight, with up to 100 different targets, complete with their image and geographical coordinates. The initial guidance is GPS based and the final guidance is completed using image-comparison software. The Spice achieved initial operational capability in 2003. The guidance kit is usually fitted to Mk 83 and Mk 84 bombs. The Spice's unique GPS and image comparison autonomous guidance system allows it to be used in a 'fire and forget' mode from single-seat fighters. The Spice is part of the arsenal of IASF F-15I and F-16C/D/I aircraft, and an enhanced version has been selected to arm the future F-35s.

F-16I serial number 854 from the 'The One' Squadron armed with a Rafael Spice guided weapon, Python 5 and AIM-120 AMRAAM missiles.
(Ofer Zidon)

Modern Israeli Air Power

Rafael Tal 1 and 2 cluster bombs
Developed in the mid-1970s, these two bombs use the same basic container canister and are operated when the canister reaches a preset altitude, after which it divides into four and the bomblets are aerodynamically armed. The Tal 1 carries 270 bomblets, while the Tal 2 carries 315 bomblets. The Tal 1 entered service in 1981 and Tal 2 followed in 1983. Both can be employed by all IASF strike aircraft.

1.3. Guidance pods

Rafael air-to-surface guidance pod
This pod is based on the American AN/AXQ-14 and is used as weapon datalink. It simultaneously transmits video from the weapon's sensors to the control aircraft and transmits commands from the control aircraft to the weapon. The pod contains two separate transmit antennas in the forward and aft radomes. The pod is used for guidance of the Popeye and Delilah missiles and is carried by all IASF F-15 and F-16C/D/I aircraft.

The Rafael air-to-surface guidance pod is visible under the centreline of this F-16I, also armed with AMRAAM, Python 4/5 and Delilah missiles. (Ofer Zidon)

Rafael Litening | Bareket (emerald stone)
The Litening pod is used for day/night navigation and laser designation and targeting. A high-resolution sensor enables pilots to reliably identify their target and to avoid collateral damage. The pod is mounted on the right-hand chin station (5a) of the F-16C/D/I and on the dedicated targeting station (located to starboard) on the F-15I. The Litening pod provides for multiple functions including:
- Laser spot detection enabling cooperative missions with target handover
- Laser marking for cooperative missions with night vision goggles (NVGs)
- Low-level night flying and navigation
- Identification of aerial targets from beyond visual range
- Detection/recognition/identification and laser designation of surface targets

- Accurate delivery of laser-guided bombs, GPS-guided munitions, cluster and general-purpose bombs and reliable damage assessment
- The pod is composed of three sections: a forward section containing the stabilised sensor unit, a central section containing the electronics units and a rear section that contains an environmental control unit. The pod is carried by IASF F-15I and F-16C/D/I aircraft.

A close-up of the Litening pod under the intake of an F-16D. (Ofer Zidon)

1.4. Electronic warfare pods

IAI Elta EL/L-8212

This self-protection pod features sophisticated jamming technology, designed to enhance the survivability of fighters and other military aircraft by suppressing multiple threats in a dense radar-guided weapon systems environment. The pod suppresses hostile weapon systems by transmitting electronic countermeasures, and the jamming of signals classified as threats. The pod is carried by IASF F-16A/B aircraft.

F-16A serial number 111 from No. 116 'Defenders of the South' Squadron lands at Nevatim with the EL/L-8212 jamming pod under its centreline pylon. (Ofer Zidon)

IAI Elta EL/L-8222
The EL/L-8222 is an advanced electronic countermeasures pod based on the EL/L-8212 electronic warfare pod. The pod is modularly constructed for multiple-band capability and ease of maintenance. The pod is mounted on the left-hand forward AIM-7 Sparrow station of the F-15A/B/C/D, with or without the conformal fuel tank fitted.

1.5. Reconnaissance pods

Elbit/Elop Condor 2
Development of this LOROP pod began in 1998. The pod system provides simultaneous high-resolution visible and infrared reconnaissance images and can be operated at a standoff range from the target. The images are stored and transmitted via a datalink. The pod is based on the F-16's 300-US gal (1,136-litre) centreline fuel tank, fitted with two 450 x 450mm (17.7 x 17.7in) right and left windows. The fully autonomous system can operate from altitudes of up to 15,240m (50,000ft).

An F-16I from No. 107 'Orange Tail' Squadron returns from a photo-reconnaissance sortie equipped with the Condor 2 pod.
(Ofer Zidon)

Elta EL-2060/P | Gondola
Development of this synthetic aperture radar (SAR) pod began in the early 1990s. It is also based on the F-16's 300-US gal centreline fuel tank, with the additions of intakes and vents. The pod contains a highly sophisticated, all-weather, day and night, SAR with ground moving target identification (GMTI). The data is stored and transmitted via a real-time datalink. The pod is operated by IASF F-16B and F-16C/D aircraft. The pod is named Gondola in IASF service.

An F-16C from No. 101 'First Fighter' Squadron returns from a mission. The EL-2060/P Gondola pod is seen under the fuselage.
(Ofer Zidon)

Chapter 5

IAI EO/IR LOROP reconnaissance pod | **Ofir**
Based on Israeli requirements, the development of a LOROP (long-range oblique photography) sensor was undertaken by the CAI company in the US. The pod's structural design and manufacturing was undertaken by IAI's Lahav factory, based on the F-15's 600-US gal (2,271-litre) centreline fuel tank, fitted with two 590 x 580mm (232 x 228in) right and left windows. The LOROP pod was initially designed for use on Israeli RF-4E and F-15 aircraft. The fully autonomous system can operate at altitudes of up to 15,240m (50,000ft). Currently, the pod is operated by IASF F-15C/D aircraft.

A close-up of the IAI LOROP reconnaissance pod and the EL/L-8222 electronic warfare pod.
(Ofer Zidon)

Rafael Reccelite | **Ogen krav (battle anchor)**
Intended for medium-altitude work, the Reccelite is a self-contained, self-cooled multi-sensor tactical reconnaissance system, consisting of an airborne pod based on the Litening targeting and navigation pod and a ground station. It simultaneously collects infrared (IR) and visual (VIS and near IR) digital images, in accordance with an automatic mission plan and/or manual operation. The images are transmitted to the ground station via a datalink.

An F-16I with the Reccelite pod under the right-hand chin station and a Litening pod on the left-hand station.
(Sariel Stiller)

The pod is mounted on the right-hand chin station (5a) of the F-16I aircraft. The forward section contains the day/night camera colour sensor. The centre section contains the datalink antenna for real-time image transfer.

1.6. Anti-tank missiles

Rafael Spike | Machtselet (mat)

A fourth-generation, long-range anti-tank missile, the Spike is a 'fire and forget' weapon with lock-on before launch capability and an autonomous guidance system. The missile is equipped with an imaging infrared seeker. The Spike is capable of flying a 'top attack' profile using a 'fire, observe and update' guidance method. The missiles can be installed in pods of up to four, which can be carried by IASF AH-1 helicopters.

AH-1F serial number 366 armed with the Machtselet system. The missile is located under the stub wing while the communication pod is located under the chin of the helicopter.
(Ofer Zidon)

1.7. Ehud instrumentation pod

The Ehud pod is built on the basis of an AIM-9 Sidewinder missile body. It measures the fighter's flight data which are transmitted to ground stations for the computing of spatial positioning of all pod-carrying aircraft on the air combat manoeuvring range. The data transmitted includes aircraft altitude, velocity and pressure data, and missile-firing data. The pod is part of the IASF's training debriefing system that allows pilots to see their actions in graphical form on a wide screen. The Ehud pod is used by all IASF combat aircraft.

Chapter 5

F-16C serial number 558 from No. 101 'First Fighter' Squadron lands after an air-to-air training sortie. The Ehud pod is attached to the missile rail on the tip of the port wing.
(Ofer Zidon)

2. US-made weapons

Besides the Israeli-made weapons described above, the IASF also uses American-made weapons in the form of general-purpose bombs, unguided bombs and some advanced guided weapons including high-penetration munitions.

2.1. Air-to-air missiles

AIM-7 Sparrow

The AIM-7 Sparrow is a medium-range air-to-air missile that employs semi-active radar homing (SARH) guidance. Although it is being phased out in favour of the more advanced AIM-120 AMRAAM, the AIM-7 Sparrow remains in IASF service. Improved versions of the AIM-7 were developed in the 1970s in an attempt to address the weapon's limitations. The AIM-7F, which entered service in 1976, has a dual-stage rocket motor for increased range, solid-state electronics for greatly improved reliability, and a larger warhead. The most common version of the Sparrow today, the AIM-7M, entered service in 1982 and featured a new inverse monopulse seeker, active radar fuse, digital controls, improved resistance to ECM, and superior low-altitude performance. Both AIM-7F and M versions are in IASF use and are carried by the F-15 family.

Despite the introduction of the AMRAAM, the AIM-7 remains in the IASF inventory.
(Ofer Zidon)

79

AIM-9M Sidewinder

The Raytheon AIM-9 Sidewinder is an infrared-homing, short-range, air-to-air missile. The AIM-9M version has the all-aspect capability of its predecessor AIM-9L model while providing all-around higher performance. The M model has improved capability against infrared countermeasures, enhanced background discrimination capability, and a reduced-smoke rocket motor. These modifications increase its ability to locate and lock-on to a target and decrease the chance of missile detection.

An F-16C in air-to-air configuration with AIM-9Ms on the wingtip stations and Python 5s on the outboard underwing pylons.
(Ofer Zidon)

AIM-120 AMRAAM

The successor to the AIM-7, the Raytheon AIM-120 Advanced Medium-Range Air-to-Air Missile (AMRAAM) provides an all-weather, beyond visual range (BVR) capability. Compared to the Sparrow, the AMRAAM's improved capabilities include increased speed, reduced size and lighter weight. It also has much improved capabilities against low-altitude targets. The AIM-120 incorporates a datalink to guide the missile to a point where its active radar is switched on for the terminal intercept of the target ('fire and forget') . An inertial reference unit and micro-computer system makes the missile less dependent upon the fire-control system of the aircraft.

A live AIM-120 AMRAAM carried on the wingtip station of an F-16I.
(Ofer Zidon)

The 'fire and forget' mode allows the aircrew to aim and fire several missiles simultaneously at multiple targets and perform evasive manoeuvres while the missiles guide themselves to the targets.

The missile also features a 'home on jamming' capability, which allows it switch over from active radar homing to passive homing – homing on jamming signals from the target aircraft. Software on board the missile allows it to detect if it is being jammed, and guide on its target using the appropriate guidance system.

Deliveries of the AIM-120C variant in use with the IASF began in 1996. The C-variant has been steadily upgraded since it was introduced, including an improved fuse (Target Detection Device), improvements in homing, and extended range.

2.2 Precision-guided bombs

GBU-15

The GBU-15 is an un-powered glide weapon. The weapon consists of five modular components that are attached to either a general-purpose 2,000lb (907lb) Mk 84 bomb or a penetrating-warhead BLU-109 bomb. The components comprise a forward guidance section, warhead adapter section, control module, airfoil components and a weapon datalink.

The guidance section is attached to the nose of the weapon and contains a daytime TV guidance or night/adverse-weather infrared system. The datalink enables the weapons systems operator on board the launch aircraft to guide the bomb to its target.

The weapon's control surfaces allow the GBU-15 to manoeuvre towards its target. The GBU-15 may be used in either a direct or an indirect attack. In a direct attack, the operator locks the weapon on to a selected target before launch and the weapon automatically guides itself to the target, enabling the launch aircraft to leave the area. In an indirect attack, the weapon is guided by remote control after launch. The operator launches the weapon and searches for the target. Once the target is acquired, the weapon can be locked on to the target or manually guided towards it.

A rare photo of a fully armed F-16I shows the Sufa armed with a GBU-15 glide bomb, Python 5 and AIM-120. An additional photo of the GBU-15 may be found on p179. (Ofer Zidon)

GBU-28

The GBU-28 'bunker buster' is a 5,000lb (2,268kg) laser-guided bomb. It was designed, manufactured and deployed in less than three weeks by the US Air Force during Operation Desert Storm in 1991, due to an urgent need to penetrate hardened Iraqi command centres located deep underground. The Enhanced GBU-28 augments the laser-guidance with an inertial navigation system (INS) and GPS guidance.

The GBU-28 contains 286kg (630lb) of high explosive. The operator illuminates the target with a laser designator and the munition guides itself to the spot of laser light reflected from the target. During tests, the GBU-28 proved itself capable of penetrating over 30m (100ft) of earth or 6m (20ft) of solid concrete.

GBU-31/32/38 Joint Direct Attack Munition

The Joint Direct Attack Munition (JDAM) is a free-fall GPS-guided bomb. The target coordinates are loaded into the guidance system of the bomb prior to launch. The JDAM's range is derived from the altitude and trajectory of the launch aircraft, with a maximum range of 30km (18.6 miles). Israeli JDAM kits are installed on the 2,000lb (907kg) Mk 84 or BLU-109 bomb (under the designation GBU-31), the 1,000lb (454kg) Mk 83 bomb (GBU-32), and the 500lb (227kg) Mk 82 bomb (GBU-38). The JDAM kit enhances the accuracy of the bombs and allows strike missions at night or under adverse weather conditions. Weapon manoeuvrability is gained using fixed aerodynamic surfaces (mid-body strakes) attached to the bomb body.

An F-16D from No. 105 'Scorpion' Squadron takes off armed with a GBU-31 JDAM. (Ofer Zidon)

GBU-39 Small Diameter Bomb

The GBU-39 Small Diameter Bomb (SDB) is a 250lb (113kg) precision-guided glide bomb, carried in clusters of four using a BRU-61 rack attached to an aircraft's hardpoint. The bomb is GPS guided with a range of 110km (68 miles) and a 206lb (93kg) warhead with high-penetration ability. The SDB enhance the effectiveness of bombing missions in three major parameters: improved precision using GPS guidance, standoff ability with long range, and the large number of bombs that can be carried on board a single aircraft – four SDBs can replace a single 1,000lb Mk 83 bomb. Israel was the initial international customer for the SDB.

Chapter 5

A GBU-39 on display with its wings deployed. Seen on the left is the BRU-61 rack with four SDBs mounted.
(Ofer Zidon)

Paveway II series

The Paveway series of laser-guided bombs was developed by Texas Instruments from 1964. Paveway kits attach to a variety of warheads, and consist of a semi-active laser seeker and a computer control group. The kit includes front control canards and rear wings for stability. The weapon guides on reflected laser energy: the seeker detects the reflected light of the designating laser, and actuates the canards to guide the bomb towards the designated point.

The improved Paveway II has a simplified, more reliable seeker and pop-out rear wings to improve glide performance. The Paveway II series of bombs has a similar range of over 15km (9.3 miles) and includes:
- GBU-10 Paveway II – Mk 84 or BLU-109 2,000lb (907kg) bomb
- GBU-12 Paveway II – Mk 82 500lb (227kg) bomb
- GBU-16 Paveway II – Mk 83 1,000lb (454kg) bomb

2.3. Anti-tank missiles

AGM-114 Hellfire

The AGM-114 Hellfire is an air-to-surface missile developed primarily for anti-armour use. Developed as the Helicopter-Launched, Fire and Forget Missile (providing the Hellfire acronym), the AGM-114 has a multi-mission, multi-target precision-strike capability.

Most variants of the Hellfire use laser guidance, although the AGM-114L (Longbow Hellfire) employs radar guidance. The latest Hellfire II, developed in the 1990s, is a modular missile system with several variants. The semi-active laser variants home in on a reflected laser beam aimed at the target. Laser guidance can be provided either from the launcher, such as the nose-mounted optronics of the AH-64, other airborne target designators or from ground-based observers. The latter two options allow the launch helicopter to break line of sight with the target and seek cover.

The AGM-114L is equipped with a millimetre wave (MMW) radar seeker, and requires no further guidance after launch, even being able to lock on to its target after launch.

While the AH-64 is theoretically capable of carrying up to 16 Hellfires, a smaller number of missiles is usually chosen for the longer-range missions typical of the IASF fleet.
(Ofer Zidon)

The Hellfire II can also be employed in adverse weather and despite battlefield obscurants such as smoke and fog. Each Hellfire missile weighs 47kg (106lb), including the 9kg (20lb) warhead, and has a range of 8km (5 miles). The AH-64 can carry up to 16 Hellfire missiles simultaneously.

BGM-71 TOW

The BGM-71 TOW (Tube-launched, Optically tracked, Wire-guided) heavy anti-tank missile is produced by Raytheon. The weapon is used in anti-armour, anti-bunker, anti-fortification and anti-amphibious landing roles and is the primary armament of the missile-armed AH-1 in Israeli service.

The TOW guidance system employs an optical sensor on the launch aircraft that continuously monitors the position of a light source on the missile relative to the line of sight. The trajectory of the missile is corrected using electrical signals generated by the optical sensor and passed down two wires to command the control surface actuators. The TOW missile was continually upgraded, with an improved TOW missile (ITOW) appearing in 1978 that had a new warhead triggered by a nose probe, which was extended after launch for improved armour penetration. The TOW 2 featured a larger warhead with a slightly extended nose probe, improved guidance and an increased-thrust motor.

Launchers allow the AH-1 to carry four TOW missiles per pylon, although a two-missile loadout per pylon is more common in IASF service. A Telescopic Sight Unit (TSU) provides targeting and guidance and has been upgraded with the Laser-Augmented Airborne TOW (LAAT), a day/night rangefinder, and Cobra Night Imaging Thermal Equipment (C-NITE), a thermal imaging/forward-looking infrared system.

Photographs of the BGM-71 can be found on pp20, 65, 97 and 233.

HATZERIM AIR BASE

Brief history

Hatzerim is located in the south of Israel near the city of Beer Sheba. With the bases of the British mandate era all located in the north and centre of Israel, the IDF/AF required a base close to the active southern border with Egypt. Hatzerim, which was the first Israeli-built base, was activated in August 1966 to host the Air Force Flight School units transferred from Tel Nof in October 1966. The new base very soon saw operational activities, when Flight School Magisters participated in the June 1967 War, performing close air support missions on the Jordanian front. Flight School instructors flew the Magisters in combat, some being killed when their aircraft, not equipped with ejection seats, were hit by ground fire.

On 17 June 1969 the Skyhawk-equipped No. 102 'Flying Tigers' Squadron was transferred from Hatzor to become the first operational squadron at Hatzerim. Two months later, No. 123 'Southern Bell' Squadron moved in from Tel Nof with its Bell 205 helicopters. When the squadron converted to the UH-60, its name was changed to 'Desert Owls'. On 1 December 1971 Hatzerim joined the F-4 community with the activation of No. 107 'Orange Tail' Squadron.

During the October 1973 War the three operational squadrons took part in combat operations. The Flight School's Magisters did not participate in combat missions as a result of the negative experiences of their participation in the June 1967 War.

A fourth squadron joined Hatzerim in August 1978. This was No. 192 'Hawkeye' Squadron operating four E-2C Hawkeye command and control aircraft. The squadron was disbanded in 1994 and three of its aircraft were sold to Mexico. The latest addition to the base's order of battle is No. 69 'Hammers' Squadron, which transferred its F-4s from Ramat David on 4 June 1991. The squadron was reactivated as an F-15I operator on 19 January 1998. During the last two decades the base has taken part in countless operational activities, together with the daily operations of the Flight School. At present, Hatzerim hosts two of the most advanced fighters in the IASF inventory – the F-15I and F-16I.

No. 69 'Hammers' Squadron

The 'Hammers' Squadron was formed on 15 July 1948 and initially operated three B-17 bombers that had been flown to Israel from Czechoslovakia. On their ferry flight they bombed targets in Cairo, Egypt before landing at Tel Nof. The squadron subsequently operated from Ramat David. The 'Hammers' flew 200 operational sorties during the War of Independence, including high-altitude bombing of Egyptian targets in Sinai and Gaza. When the war was over the squadron was transferred to Tel Nof. The early 1950s saw the 'Hammers' fly photography and patrol missions, exploiting the B-17's high-altitude and long-range capabilities.

In December 1954 the squadron was deactivated and its bombers were transferred to the 'Flying Elephant' heavy transport squadron, but the 'Hammers' were reactivated again on 1 May 1956 in the wake of the Sinai Campaign that began on 29 October of that year. During the Sinai Campaign the squadron flew eight missions, releasing seven tons of bombs. The B-17 era in Israel formally concluded on 1 April 1957 with the closure of the squadron and the sale of the B-17s.

On 1 November 1969 the 'Hammers' Squadron was reactivated in its new incarnation, as the second F-4 squadron of the IDF/AF. The 'Hammers' was re-formed at Hatzor – alongside its older sister unit, No. 201 'The One' Squadron – but was transferred to Ramat David on 15 November 1969, where it served until 1991. The usual challenge of forming a new squadron with new aircraft was magnified by the fact that this took place during the War of Attrition. The new squadron was thrown into battle merely three weeks after recommissioning. On 11 November 1969, 'Hammers' aircrew claimed their first MiG kill with the F-4, flying a Phantom that belonged to 'The One' Squadron. The first 'full' 'Hammers' kill was claimed on 8 February 1970, and was also the first to be claimed using the cannon. On 28 November 1969 the squadron flew its first air strike, attacking an Egyptian SA-2 SAM site. The squadron, together with 'The One' Squadron, carried the heavy burden of air strikes against the expanding Egyptian SAM network between Cairo and the Suez Canal. Heavy losses were sustained in the process – three of the squadron's Phantoms were lost and six crewmembers became PoWs. During an air strike on 18 July 1970 both squadron commanders were hit by anti-aircraft missiles. The commander of 'The One' Squadron, Shemuel Chetz, was killed and the commander of the 'Hammers', Avihu Ben Nun, performed successful emergency landing. On 30 July 1970 the squadron scored two kills during an aerial battle against Soviet pilots flying Egyptian MiGs.

The squadron flew 850 operational sorties during the October 1973 War. The 'Hammers' claimed 12 kills over Syria and Egypt and lost nine Phantoms, with four aircrew killed and eight captured. The rest of the crews ejected and were rescued. During the conflict the 'Hammers' were reinforced with 20 more Phantoms that arrived straight from US Air Forces in Europe squadrons.

In March 1978 the squadron participated in Operation Litani in Lebanon, flying 50 operational sorties comprising day and night air strikes using AGM-65 Maverick missiles and bombs, reconnaissance sorties and interception patrols. During the First Lebanon War in June 1982 the squadron took major part in the attack against Syrian SAMs in Lebanon, with its Phantoms leading the first wave of the attack. On 16 October 1986, during a mission over Lebanon, one of the squadron's Phantoms was lost when one of the bombs it was carrying exploded. The crew ejected and the pilot was safely rescued on board the skids of an AH-1. The navigator, Ron Arad, was captured. For some years he was transferred For some years he was transferred between terrorist groups in Lebanon until he was finally posted as MIA, supposedly in Iran.

In 1991 the IDF/AF decided to purchase an Israeli version of the F-15E, the F-15I Ra'am. This strike version of the F-15 could carry a payload of over seven tons and it

Chapter 6

was natural to assign the new aircraft to the 'Hammers' Squadron, with its heritage as a heavy bomber operator. On 19 January 1998 the squadron was reactivated and received its first pair of F-15Is.

In its latest incarnation as an F-15I operator, No. 69 'Hammers' Squadron is considered to be the spearhead of the IASF, and the long-range strategic arm of Israel. As such, it can be assumed that it probably took part in long-range operations against high-value targets, such as the attacks on weapons convoys in Syria, the attack on the alleged nuclear facility in Syria's eastern desert, and more.

F-15I serial number 252 is depicted armed with laser-guided bombs in the 'heavy' air-to-ground configuration. (Ofer Zidon)

F-15I serial number 263 awaits take-off clearance from the Hatzerim control tower. (Ofer Zidon)

Modern Israeli Air Power

F-15I serial number 215. The aircraft is flying with no external load to maximise manoeuvrability during an aerobatic display.
(Ofer Zidon)

F-15I tail number 246 taxis from its HAS and out to the runway during a deployment to Ovda. Note the SATCOM fairing aft of the cockpit and the 500lb (227kg) practice bomb under the wing.
(Ofer Zidon)

Chapter 6

No. 102 'Flying Tigers' Squadron

On 29 December 1967, following the arrival by sea of the first batch of Israeli A-4 Skyhawks, a new squadron was commissioned to receive the new attack aircraft. This was No. 102 'Flying Tigers' Squadron, which became the second squadron to operate the A-4 after No. 109 'Valley' Squadron. At first, the 'Flying Tigers' pilots flew within the 'Valley' Squadron. The 'Flying Tigers' finally received their first batch of aircraft on 13 June 1968 and began to fly these from Hatzor. A year later, in June 1969, the squadron was transferred to Hatzerim and became the new base's first combat squadron. During the

A mix of A-4N and TA-4J aircraft await clearance to take off from Hatzerim.
(Ofer Zidon)

A-4N serial number 303 taxis from the line to Hatzerim's runway.
(Ofer Zidon)

89

Modern Israeli Air Power

TA-4J serial number 730 at Hatzerim. Note the 'Improved Ahit' inscription behind the national insignia on the intake and the winged tiger tail art. (Ofer Zidon)

TA-4H serial number 544 takes off from Hatzerim. This aircraft is the world's oldest flying A-4. (Ofer Zidon)

War of Attrition the squadron flew attack missions against all fronts – Egypt, Syria, Jordan and Lebanon. In 1972 the A-4 assumed the advanced training phase of the flight course within the Air Force Flight School, a role it fulfils to this day.

During the October 1973 War the squadron flew 1,000 operational sorties, primarily ground-attack missions against Syrian and Egyptian ground forces, anti-aircraft batteries and armoured forces. The squadron suffered severe losses: 17 of its jet were destroyed, mostly by AA fire, with seven pilots killed and five captured to become PoWs. The attrition war on the Syrian front which followed the end of the October 1973 War saw the squadron involved in numerous attack sorties in Syria, Mount Hermon and over Lebanon.

The next phase in the squadron's development began in May 1976, with the arrival of the advanced A-4E variant. These aircraft were transferred from No. 140 'Golden Eagle' Squadron and were fitted with Israeli-made avionics, including the Kristal head-up display and mission computers.

In 1981 the 'Flying Tigers' took charge of the operational training course for new pilots, after the previous operator of the OTU, No. 140 'Golden Eagle' Squadron, ceased flying when it was transferred to Ramon. The 'Flying Tigers' continues to run operational training courses to this day, together with the Nevatim-based No. 116 'Defenders of the South' Squadron.

During the 1982 First Lebanon War the squadron's A-4s flew 177 operational sorties, mainly ground-attack. The Skyhawks also flew electronic warfare missions, making use of the aircraft's long endurance.

On 16 March 1986 the squadron transferred to the latest type of the Skyhawk – the A-4N – that it still flies today, alongside the two-seat TA-4J. As such, the 'Flying Tigers' is the final Israeli Air Force Skyhawk operator.

In 2005 the squadron's aircraft underwent an improvement programme to extend their service life until 2015, when they are expected to be replaced by the M-346 Master.

No. 107 'Orange Tail' Squadron

On 25 January 1953, No. 107 'Orange Tail' Squadron was established as a fighter-bomber unit flying the Spitfire from Ramat David. It was the IDF/AF's third Spitfire squadron and was responsible for operational training courses for new graduates from the Flight School. In December 1953 the squadron became a reserve unit and four months later it converted to the P-51. Two years later, in June 1956, the 'Orange Tail'

A line of F-16Is awaits their crews. Note the absence of CFTs on the nearest aircraft. (Ofer Zidon)

Squadron returned to its origins when it exchanged its Mustangs for the Spitfires of No. 105 'Scorpion' Squadron. After the sale of the Spitfires to Burma the squadron was disbanded in September 1956.

In February 1962 the squadron was reactivated at Ramat David. No. 117 'First Jet' Squadron was about to receive the new Mirage III fighters and handed over its Meteors to the 'Orange Tail'. After flying the Meteors for two years, the decision was made to retire the single-seat Meteors and transfer the two-seat Meteors to the Vautour-equipped No. 110 'Knights of the North' Squadron to serve as trainers. With that decision the 'Orange Tail' Squadron was deactivated once again.

On 15 September 1965 the squadron was activated for the third time, following the purchase of additional Ouragans between late 1964 and early 1965. In its new incarnation the squadron was focused on ground attack. During the June 1967 War the squadron was deployed to Lod near Tel Aviv, in order that the relatively short-ranged Ouragans could attack airfields and radar installations in Egypt and Jordan. Other targets included ground forces, AA batteries and convoys on all fronts. When the war ended the squadron transferred its Ouragans to No. 113 'Hornet' Squadron and was disbanded again.

On 1 December 1971 the squadron was recommissioned at Hatzerim, flying the fourth aircraft type in its history, the new and most advanced fighter of the IDF/AF – the F-4. The 'Orange Tail' Squadron became the third Phantom squadron within the IDF/AF. On 2 January 1973 the squadron was credited with its first Syrian MiG-21 kill over Lebanon. On 13 September 1973, a photo-reconnaissance sortie developed into a large-scale aerial battle over Syria with 12 MiGs claimed as shot down, three of them by 'Orange Tail' Phantoms.

The squadron flew 752 operational sorties during the October 1973 War, in a mix of interception and ground-attack missions on the Egyptian and Syrian fronts. On the first day of the war a pair of the squadron's Phantoms was on a QRA mission at Ofir, the most southerly base in Sinai, when this came under attack by Egyptian MiGs and the pair claimed seven enemy fighters as shot down. The area was also a target for an Egyptian Army commando battalion flown in by Mil Mi-8 helicopters. Seven of the helicopters were claimed as shot down and the assault was severely curtailed. On the fourth day of the war, eight of the squadron's Phantoms took part in the attack on the Syrian Army HQ in Damascus. Due to bad weather conditions the Phantoms could not

F-16I serial number 891 armed with a Rafael Spice guided munition. The position of the CFTs is shown to advantage in this view.
(Ofer Zidon)

Chapter 6

F-16I serial number 879 takes off from Ovda during a deployment to the southern base.
(Ofer Zidon)

Providing a clear view of the SATCOM fairing on the spine, F-16I serial number 827 lands at Ovda during a squadron deployment.
(Ofer Zidon)

complete the attack and on their way back, fully armed, they attacked a Syrian tank concentration. The 50 tons of bombs were claimed to be a major factor in blunting the Syrian advance in the south of the Golan Heights. During the war the squadron claimed 28 enemy aircraft as shot down and lost four of its own Phantoms – three to enemy AA fire and one to a Mirage III in a 'friendly fire' incident. All crews were rescued safely.

Following the October 1973 War the squadron participated in countless operations, including Operation Litani in 1978, the First Lebanon War in 1982, Operation Grapes of Wrath in 1996, and air strikes in Lebanon.

On 5 July 2006 the 'Orange Tail' Squadron was reactivated again, as the third F-16I unit in the IASF. Preserving the legacy of its Phantom operations, the squadron operates from Hatzerim, while the other three F-16I squadrons fly from Ramon.

No. 123 'Desert Owls' Squadron

On 10 March 1965 the second helicopter squadron of the IDF/AF was activated, as No. 123 'Southern Bell' Squadron. Its primary mission was the training of new helicopter pilots and the first helicopter pilots' course began on 15 April 1965, using 13 Bell 47Gs and two Sikorsky S-58s loaned from No. 124 'Rolling Sword' Squadron. The course attendants were mix of experienced fighter pilots and new flying course graduates. The course lasted for seven months during which the cadets started flying on the Bell 47 and later moved on to the S-58. When the Air Force Flight School moved to Hatzerim the squadron remained in Tel Nof.

The squadron converted to the Bell 205 in December 1967 and the new helicopter took the place of the Bell 47 and S-58 for the training of new pilots. On August 1969 the squadron was transferred to Hatzerim, from where it has operated ever since.

During the October 1973 War both helicopter squadrons were assigned missions of liaison and transport and the evacuation of wounded soldiers from the battlefield. Some 890 wounded were evacuated using the services of the helicopter squadrons. Lessons learned during the war led to the replacement of the Bell 205 with the new and more powerful twin-engine Bell 212. The new helicopter arrived with the squadron in 1978.

UH-60L Yanshuf 3 serial number 587 parked on the squadron apron at Hatzerim. (Ofer Zidon)

Chapter 6

UH-60A Yanshuf 1 serial number 741 during an assault transport demonstration at Hatzerim.
(Ofer Zidon)

UH-60L Yanshuf 3 serial number 555. This is one of the newly built helicopters that arrived in Israel already wearing the desert camouflage scheme.
(Ofer Zidon)

Modern Israeli Air Power

Besides the training of new pilots, the squadron participated in numerous operational roles, including the assault transportation of troops in and out of the battlefield, search and rescue, electronic warfare missions and others.

In 2002 the squadron received the new UH-60, a vast improvement over the veteran Bell 212. The helicopter was called Yanshuf (owl) in IASF service and the squadron name was thus changed from 'Southern Bell' to 'Desert Owls'. During the 2006 Second Lebanon War the squadron flew 66 operational sorties and evacuated 43 wounded soldiers from Lebanon and Gaza.

Air Force Academy

Alongside the F-15I, F-16I and UH-60 squadrons, the volume of air traffic related to the Air Force Academy makes Hatzerim one of the busiest air bases in Israel. The Academy is divided into fixed- and rotary-wing training divisions. An air cadet has to pass through a number of flying phases, that include screening, basic training and advanced training. Alongside these flying phases the future pilot must pass through basic military training, and streaming flights aimed at directing him to one of three streams: fast jet, transport or rotary wing, and pilot or navigator, as well as two terms of academic studies at Ben Gurion University in the Negev.

The course is divided into terms, each lasting six months. The preparation term includes the screening flights (flown on the Grob 120) and basic military training. The basic term consists of basic flying training (T-6) and the officer's course. Next is the initial term that sees the student learns initial flying skills (T-6 or Bell 206). There follows two terms of academic studies prior to the advanced phase, which comprises advanced flying training (A-4, AH-1, UH-60). Throughout a total of six terms (across three years) in which the cadets fly around 250 hours with 50 simulator hours.

A. Fixed-wing training

The Academy's syllabus includes a short phase of screening, testing the candidate's aptitude for flying, followed by basic and advanced flying training phases. In the past, the screening phase was completed using the Super Cub, while the two later phases until recently flown with all-jet equipment. The Magister was used for both basic and

T-6 serial number 497 from the Air Force Academy's basic training squadron. Note the Aerobatic Team emblem below the front cockpit.
(Ofer Zidon)

Chapter 6

A formation of three Grob 120s over Hatzerim. The Grob 120 is owned by Elbit Systems (hence the civil registration in the 4X-range), which sells flying hours to the IASF.
(Ofer Zidon)

T-6 serial numbers 414 and 483 from the Air Force Academy's basic training squadron. The aircraft are displayed to the public during the graduation ceremony for new pilots held twice a year at Hatzerim.
(Ofer Zidon)

97

advanced training following its introduction in 1960, while the A-4 assumed responsibility for the advanced training role from 1972. Over the years the Magister underwent a comprehensive upgrade programme by IAI, which resulted in the Tzukit. More than 80 upgraded Tzukit airframes were delivered to the IDF/AF between 1981 and 1986.

The modernisation of the training fleet began in 2002, with Elbit Systems winning a contract to sell the Academy flying hours on the Grob 120 for the screening phase. The second phase of modernisation took place in July 2009, with the arrival of the first four of 20 Beechcraft T-6A Texan II trainers.

The A-4 advanced training fleet underwent an upgrade programme beginning in 2005, and aimed at extending its service life for another 10 years, until 2015, when it will be replaced by the Alenia Aermacchi M-346.

B. Rotary-wing training

The Air Force Academy's rotary-wing training syllabus includes the same phases as the fixed-wing line: primary training in the Bell 206 and advanced training in either the AH-1, for advanced attack helicopter training, or the UH-60 for advanced assault helicopter training. Air cadets unsuited for fast-jet flying continue their basic training phase on the Bell 206, which now serves exclusively as a trainer. Assault training makes use of UH-60s shared with No. 123 'Desert Owl' Squadron at Hatzerim, while attack helicopter training involves a handful of AH-1s on strength with the Academy, and with operational equipment deleted.

Bell 206 serial number 012 from the Academy's rotary-wing basic training squadron. After its retirement from front-line service, the Bell 206 is now painted in a high-visibility colour scheme.
(Ofer Zidon)

Chapter 6

UH-60L Yanshuf 3 serial number 582 from No. 123 'Desert Owls' Squadron. The unit is responsible for the advanced training assault helicopter stream, as well as operational training for graduate assault helicopter pilots.
(Ofer Zidon)

AH-1E serial numbers 652 (airborne) and 658, from the Academy's rotary-wing advanced training attack helicopter stream.
(Ofer Zidon)

C. Operational training

Graduation from the Air Force Academy, with the long prized pilot's wings, is only the first step on the long road to becoming a combat pilot. The next step is operational training. After the Academy has taught the graduates to fly, the operational training will teach them how to fight in an aircraft. Those graduates that are assigned to the fly-by-wire F-16 squadrons will begin a 'Netz Year' at Nevatim, which hosts the IASF's two F-16A/B Netz squadrons. 'Netz Year' comprises a year of postgraduate training, divided into a six-month Operational Training Unit (OTU) course and another six-month Advanced Operational Training Unit (AOTU) course. After the 'Netz Year' the fighter pilots are assigned to their operational squadrons, in which they will begin their third and final step – conversion to the squadron's aircraft type and daily operational training for the squadron's specific missions. Graduates that are assigned to the F-15 Baz or Ra'am fleets will likely spend their OTU and AOTU training flying the A-4 with No. 102 'Flying Tigers' Squadron at Hatzerim.

In the helicopter stream, No. 123 'Desert Owls' Squadron is also responsible for operational training of assault helicopter graduates.

A-4N serial number 342 from No. 102 'Flying Tigers' Squadron. The squadron is responsible for the advanced training phase and for operational training of graduates.
(Ofer Zidon)

Chapter 6

F-16B serial number 001 from No. 140 'Golden Eagle' Squadron at Nevatim. The unit is responsible for the operational training course for young graduate pilots and therefore operates a considerable number of two-seat F-16s. The operational training course takes place in the first half of the 'Netz Year', following graduation from the Academy.
(Ofer Zidon)

F-16A serial number 135 from No. 116 'Defenders of the South' Squadron at Nevatim. The squadron is responsible for the advanced operational training course for young graduate pilots, which takes place during the second half of 'Netz Year'.
(Ofer Zidon)

101

Modern Israeli Air Power

Aerobatic Team

The Air Force Academy performed aerobatic display flights using various types on an ad hoc basis from 1950. A 1954 accident at Ramat David led to the decision to form an official aerobatic team, with stricter rules for improved safety. In the same year, the Aerobatic Team of the Air Force Academy at Tel Nof made its debut, flying Harvards. In the early 1960s the team re-equipped with the new Magister jet trainer.

Today, the Aerobatic Team is manned by Academy instructors. The team flies only three or four displays a year, including graduation ceremonies, Israeli Independence Day and Israeli Air Force Day. Usually, the pilots assemble two to three weeks prior to each event and train basic aerobatic manoeuvres. Overall, their display is simple when compared to specialised aerobatic display teams. The display is based on a four-ship formation, which completes two or three two-aircraft crossovers. After the display, the pilots return to their regular duties. When the Academy transferred from the Magister to the T-6, the Aerobatic Team also transferred to the new aircraft in June 2010.

The T-6s of the Aerobatic Team perform a display. Note the red undersides of the wings. (Ofer Zidon)

Chapter 7

HATZOR AIR BASE

Brief history

Hatzor air base is the Israeli reincarnation of RAF Qastina, which was built during the British mandate in Palestine between 1923 and 1948. The base saw operations by Handley Page Halifax and Douglas Dakota squadrons of the RAF. With the British withdrawal from Palestine in 1948, soldiers from the IDF's Givati infantry brigade took control of the base. It was named Shmuel airfield after Sam Pomerantz, one of the early IDF/AF fighter pilots killed during the transfer of Spitfire fighters from Czechoslovakia to Israel in December 1948. A few months later No. 101 'First Fighter' Squadron relocated to the base from Herzliya airfield.

In 1949 the 'First Fighter' Squadron was transferred to Ramat David and Hatzor was handed over to the artillery corps. It was returned to the Air Force two years later.

The second half of the 1950s saw Hatzor become the cornerstone of the French era in the ID/AF. No. 113 'Hornet' Squadron was activated in 1955 flying the latest IDF/AF jet fighter – the French Ouragan. The 'First Fighter' Squadron returned to Hatzor and was reactivated with the arrival of the new Mystère IVA, to become once again the cutting-edge interceptor squadron of the IDF/AF. Between them, the two squadrons flew 323 operational sorties during the Sinai Campaign in October 1956, including assisting in the capture of the Egyptian Navy destroyer *Ibrahim el-Awal*.

Hatzor once again became home to the most advanced jet in Israel when No. 105 'Scorpion' Squadron was reformed in 1958 as the third jet squadron at the base, flying the Super Mystère B2. In April 1962 the 'First Fighter' Squadron was re-equipped with the supersonic Mirage IIICJ interceptor and Hatzor again accommodated the latest and most advanced type in the IDF/AF inventory.

During the June 1967 War the Hatzor squadrons flew 1,294 operational sorties that included attacks on enemy airfields and anti-aircraft batteries, close air support to ground forces, and top cover.

On 17 August 1969 a new squadron was established in Hatzor. This was No. 201 'The One' squadron, the first F-4 squadron within the IDF/AF. The new squadron joined the older squadrons in Hatzor to generate hundreds of combat sorties during the War of Attrition from 1969–70, with 'The One' Squadron taking the lead for long-range attacks deep into Egypt. By the end of the War of Attrition the base had posted claims for 42.5 kills of enemy aircraft. During that time the then recently established IAI upgraded No. 105 'Scorpion' Squadron's Super Mystère B2s to Sa'ar (tempest) standard, with the integration of the American J52 engine (as used in the A-4) into the French airframe.

The October 1973 War saw the base's four squadrons fly a total of 3,137 ground-attack and interception sorties. No. 101 'First Fighter' Squadron flew mainly air superiority sorties with the IAI Nesher (eagle) and Mirage III, claiming 56 enemy aircraft. The squadron commander, Lt Col Avi Lanir, was shot down over Syria, captured and later died – Israeli accounts report that he was tortured to death. No. 201 'The One' Squadron flew 758 sorties, mostly ground attacks against Egyptian and Syrian targets, including the Syrian Army HQ in Damascus. The squadron claimed 32 enemy aircraft as shot down and lost 14 of its own aircraft, together with seven crewmembers killed and 14 captured. No. 105 'Scorpion' Squadron flew 887 ground-attack sorties in its new Sa'ar airframes and lost two pilots plus another shot down and captured. No. 113 'Hornet' Squadron flew 660 combat sorties with its Neshers, most of them air superiority sorties, claiming 52 enemy aircraft shot down for no loss.

Hatzor saw more changes in 1975, when No. 105 'Scorpion' Squadron transitioned from the Sa'ar to become the fifth and final F-4 squadron, and No. 101 'First Fighter' Squadron gave up its Mirage IIIs to become the first IAI Kfir squadron in the IDF/AF. In 1976 No. 113 'Hornet' Squadron received its Kfir fighters and brought an end to the French era at Hatzor.

After the peace agreement with Egypt, Israel returned the Sinai Peninsula and evacuated its air bases from the area. In 1982 No. 254 'Midland' Squadron moved to Hatzor with its Kfirs and in 1985 the squadron was disbanded together with the 'Hornet' Squadron.

In the 1982 fighting the five squadrons at Hatzor flew 1,196 combat sorties including the successful attack on the Syrian anti-aircraft array in the Bekaa Valley in Lebanon, led by F-4s from the 'Scorpion' Squadron.

The changes in the base order of battle continued into 1987 with the opening of the F-16C/D era in No. 101 'First Fighter' Squadron. In 1988 two Kfir squadrons joined Hatzor – No. 144 'Phoenix' and No. 149 'Smashing Parrot' – while 'The One' moved to Tel Nof with its F-4s. In 1991 the 'Smashing Parrot' Squadron disbanded and the 'Scorpion' Squadron was reactivated as the second F-16C/D outfit at Hatzor. In 1994 the 'Phoenix' Squadron traded its Kfirs with F-16A/Bs before the squadron was deactivated in 2005.

As of today, Hatzor supports two fighter squadrons: No. 101 'First Fighter' and No. 105 'Scorpion' Squadron. It also accommodates the 'Simulator' Squadron, equipped with F-16C/D, F-16I, F-15 Baz and F-15I Ra'am simulators, and an Iron Dome unit (equivalent to a fighter squadron).

No. 101 'First Fighter' Squadron

No. 101 'First Fighter' Squadron was established in Ekron (later named Tel Nof) in May 1948. The squadron's first aircraft were Avia S-199s – Czechoslovak-built Messerschmitt Bf 109s. The aircraft were delivered to the new state of Israel on board Curtiss C-46 Commando transports. The squadron's first operational sortie took place on 29 May 1948, when its entire fleet of four Avias took off from Ekron to strike Egyptian forces close to Ashdod, about 30km (18.6 miles) south of Tel Aviv. The sortie succeeded in surprising the Egyptian forces and halted the Egyptian column in its advance towards Tel Aviv. This first mission cost the lives of one pilot and two aircraft.

In November 1948 the squadron moved to Hatzor. The 'First Fighter' Squadron claimed 12 enemy aircraft shot down during the War of Independence. The squadron

Chapter 7

F-16C serial number 530 in air-to-air configuration with AIM-9 and Python 5 AAMs. (Ofer Zidon)

commenced additional roles during the same conflict, including close air support of ground forces, photography missions, and the training of graduates from the first two flight courses of the young IDF/AF, including an operational training course. In June 1949 the squadron was relocated to Ramat David, the ID/AF's premier combat air base. At that time the 'First Fighter' Squadron flew a mix of Spitfire Mk 9s and P-51D Mustangs. In August 1950 the squadron ceased providing the operational training course and this responsibility was handed over to new formation that later became No. 105 'Scorpion' Squadron.

In April 1956 the 'First Fighter' Squadron was re-established as a Mystère squadron. During the Sinai Campaign the squadron's pilots claimed seven Egyptian aircraft shot down, during the course of air superiority and some close air support missions.

The 'First Fighter' Squadron flew the Mystère until April 1962, when it received the Mirage III. The Mirage was the first Mach-2 jet in the IDF/AF, and the first offering guided air-to-air missile armament and true day and night capabilities.

An F-16C displays an unusual configuration of four Rafael Spice guided munitions and two AIM-9 Sidewinder AAMs. (Ofer Zidon)

105

Modern Israeli Air Power

During the June 1967 War the 'First Fighter' Squadron was the leading interceptor unit in the IDF/AF, but it also took part in Operation Moked (focus), striking Egyptian airfields and also flying many close air support missions. During the war the squadron flew 337 combat sorties, with four of its aircraft shot down and three pilots killed. The squadron was credited with shooting down 14 enemy aircraft and destroying 35 aircraft on the ground during the first day of the war.

With the arrival of the F-4 during the War of Attrition, the Mirage III squadron focused on the air superiority mission over Sinai and Egypt's Nile River delta. This long war resulted in claims for 27 Egyptian fighters shot down, with one No. 101 Squadron pilot taken prisoner and three Mirages lost in aerial combat. The squadron participated in the famous aerial combat against Egyptian MiG-21s flown by Soviet pilots on 30 July 1970, which ended with the shooting down of five of the Soviet pilots.

During the October 1973 War the squadron carried out air superiority and reconnaissance missions. The 'First Fighter' Squadron was the leading interceptor unit within the ID/AF, with its Mirage III and Nesher fighters claiming 56 enemy aircraft shot down at the cost of four aircraft lost and the squadron commander Avi Lanir shot down and killed.

In 14 April 1975 the squadron received its first Kfir fighters and a year later it was equipped with the improved Kfir C2, which participated in its first ground-attack mission on 9 November 1977. During the 1982 First Lebanon War, the Kfir-equipped squadron primarily flew close air support missions.

In December 1987 the squadron received its first F-16D Barak 2. This symbolised the return of the squadron to the upper echelons of IDF/AF operational units, after its second-line Kfir years. The squadron then flew a mix of F-16C and F-16D aircraft. This

A trio of F-16Cs prepares to take off from Hatzor. (Ofer Zidon)

106

Chapter 7

Armed with a JDAM, F-16C serial number 534 lands after an air-to-ground practice sortie. (Ofer Zidon)

was the squadron's first experience of flying a two-seat aircraft and it had to develop new methods to integrate pilot and navigator work. In 2004 the two heterogeneous Hatzor-based F-16C/D squadrons swapped aircraft and aircrew to become homogenous squadrons – the 'First Fighter' Squadron became a single-seat F-16C unit while the 'Scorpion' Squadron became a two-seat F-16D unit.

During its years operating the F-16C/D, the 'First Fighter' Squadron has participated in numerous combat operations over the Gaza Strip and Lebanon.

Armed with a Rafael Spice guided munition, F-16C serial number 557 takes off from Ovda during one of the squadron's deployments. (Ofer Zidon)

107

No. 105 'Scorpion' Squadron

The roots of No. 105 'Scorpion' Squadron lie with No. 101 'First Fighter' Squadron. On 20 August 1950 the 'First Fighter' Squadron assumed the operational training of new pilots. The outfit in charge of operational training expanded and in December 1950 it became a separate squadron flying some 20 Spitfires handed over by the 'First Fighter' Squadron, which was re-equipped with P-51Ds. In 1956 the new squadron received its first P-51Ds from No. 107 'Orange Tail' Squadron to become a fully operational combat squadron.

During the 1956 Sinai Campaign the squadron assumed close air support and ground-attack roles including the cutting of telegraph wires all over Sinai. The squadron lost three aircraft and the squadron commander Maj Moshe Tadmor was killed when his Mustang was shot down. On 20 January 1957 the squadron was disbanded.

On 20 August 1958 the 'Scorpion' Squadron was reactivated at Hatzor to be equipped with the Super Mystère B2. The new jet fighter was the first in Israeli service to fly supersonic in straight and level flight. The 'Scorpion' Squadron was the leading interceptor unit within the IDF/AF until the arrival of the Mirage III in 1962.

During the Six-Day War the squadron flew mainly air-to-ground combat sorties, claiming destruction of 55 enemy aircraft on the ground in Operation Moked on the first day of the war. Later it participated in air strikes on Egyptian airfields and facilities, close air support and some air-to-air missions, claiming five enemy aircraft in 507 combat sorties. Nine of the squadron's aircraft were shot down during the war, with six pilots killed and one taken prisoner.

F-16D serial number 687 armed with a small practice bomb and seen during final checks before taxiing to the runway. (Ofer Zidon)

Chapter 7

F-16D serial number 638 takes off armed with a JDAM underwing.
(Ofer Zidon)

F-16D serial number 676 departs Hatzor armed with a Delilah missile underwing. The missile guidance pod is on the centreline
(Ofer Zidon)

109

During the War of Attrition the squadron again flew mainly air-to-ground missions, while its Super Mystères were upgraded to Sa'ar standard.

In the October 1973 War the 'Scorpion' Squadron flew its Sa'ars in 887 combat sorties that included close air support and ground-attack missions. The squadron lost two pilots in the war and another one was shot down and captured by enemy forces.

Between 1975 and 1987 the 'Scorpion' Squadron became the IDF/AF's fifth and final F-4 squadron. It played a major role in the 9 June 1982 operation against Syrian anti-aircraft batteries in the Bekaa Valley in Lebanon, claiming destruction of 14 SAM batteries for no loss. On 11 June 1982 the last IDF/AF F-4 kill was achieved by the 'Scorpion' Squadron, which claimed a Syrian MiG-21 shot down.

The squadron was re-established on 24 December 1991, following its sister 'First Fighter' Squadron in receiving the F-16C/D. The squadron flew mixed single- and two-seat F-16s until 2004 when it swapped its single-seat F-16Cs for the F-16Ds of the 'First Fighter' Squadron, and became a homogenous F-16D squadron. During its F-16 years the squadron has carried out thousands of sorties over Gaza and Lebanon.

'Simulator' Squadron

Traditionally, simulators were used to train pilots in emergency procedures that are too dangerous to be practiced in real flying, and to practice flying in non-visual conditions (instrument, bad weather and night time). However, due to the great advance in simulator technology and computer systems, modern simulators allow pilots to practice other missions, from basic competence training to the use of advanced and expensive weapons such as laser- and TV-guided munitions. Simulator hours are therefore part of a squadron's allocated flying hours.

The next step in simulator use is the simultaneous training session in which up to eight trainees 'fly' together in multi-aircraft missions (formations of two, four or eight aircraft), sharing information and threats and practicing complex attack and defence tactics. This also allows pilots from different fleets to practice together, learning more about each other's aircraft advantages and shortcomings. Moreover, it allows pilots to train together with air control and ground units. State-of-the-art simulator technologies, with real-life scenery projected on to a wide screen, supported by fast and powerful computer systems, have advanced to a point where the pilots can hardly tell the difference between a simulator training mission and real flying.

Within the IASF, the latest development in this field took place in the October 2010 activation of a second simulator squadron, based at Hatzor. The new squadron operates simulators for the F-15 Baz/Ra'am and F-16 Barak/Sufa fighter force. Elbit Systems won the contract to develop an advanced simulator, which is in fact a group of eight F-16 simulators interconnected, to allow training of up to eight-ship formations. The new simulator array, which was scheduled to become operational in late 2012, also includes two Red Force adversary simulators. It furthermore incorporates built-in threats such as anti-aircraft missile launches.

To emphasise the importance of simulator training, the IASF High Command has decided that between five and 10 per cent of fighter pilots' training hours will be in the simulator.

Chapter 7

A view of the F-16I simulator at Hatzor.
(Ofer Zidon)

A view of the control panels of the F-16I simulator and its attendant simulator instructors.
(Ofer Zidon)

111

Chapter 8

NEVATIM AIR BASE

Brief history

Nevatim was one of the three bases built in the early 1980s after Israel's withdrawal from Sinai in the wake of the peace agreement with Egypt. Construction of the base was completed in 1983 and the first squadron to populate the new base was No. 116 Squadron. At that time it flew the A-4 and was known as the 'Flying Wing' Squadron. A year later, in 1984, another unit joined operations at the new base. This was a second A-4 operator, No. 115 'Flying Dragon' Squadron that was relocated to Nevatim from Tel Nof. On 17 July 1985, with the conversion of the Ramon-based No. 140 'Golden Eagle' squadron from the A-4 to F-16, the Nevatim-based 'Flying Wing' Squadron assumed its duties as an Operational Training Unit. From this date, new graduates of the flight course made their way to Nevatim for the OTU and Advanced OTU courses. Between 1991 and 1994 the operational training mission was divided between the two Nevatim squadrons: the 'Flying Wing' and the 'Flying Dragon' Squadrons.

The 'Flying Dragon' Squadron was disbanded on 21 July 1994, not long after flying 38 operational sorties during Operation Accountability, a week-long attack that targeted Hezbollah militants in Lebanon in July 1993. This left the 'Flying Wing' Squadron as the only operational outfit in Nevatim. At that time, with the reduction of operations in Nevatim, the IDF/AF High Command had to decide between two options – relocating the 'Flying Wing' Squadron to Hatzerim and closing the base, or initiating a comprehensive upgrade of infrastructure at Nevatim in order to support the relocation of the Heavy Transport Wing from Lod (adjacent to Ben Gurion Airport) to the Negev. The IDF/AF chose the second option in parallel with an Israeli government decision to relocate IDF and IDF/AF bases and operations from central Israel to the south. The plan began to be realised when work on the 'transport side' of Nevatim started in 2002. A relocation fly-past to mark the relocation of the Boeing 707 and C-130 fleets from Lod to Nevatim took place on 27 August 2008.

Meanwhile, other changes took place on the 'fighter side' of Nevatim. During 2003 Nevatim became the home of the IDF/AF F-16A/B operations. First, the 'Flying Wing' Squadron retired its A-4s to receive the Ramon-based 'Negev' Squadron's F-16A/Bs (the 'Negev' Squadron was to receive the F-16I in 2004), and adopted the new name 'Defenders of the South'. Then the F-16A/B-equipped 'Golden Eagle' Squadron relocated with its aircraft to Nevatim from Ramon. The two F-16 Netz squadrons continued Nevatim's historic role as a centre of operational training for new pilots, although in 2013 the fleet was consolidated into a single unit, No. 116 'Defenders of the South' Squadron.

With the relocation of the Heavy Transport Wing, Nevatim's status was upgraded from wing to base. Nevatim is today the only IASF base supporting both fighter and transport operations, dictating an ongoing effort of learning procedure, as well as construction work to ensure the effectiveness of the two forces operating side-by-side. The first base commander was a transport pilot and his deputy was a fighter pilot. Three years later the new commander came from the fighter community while his deputy came from the transport community.

In the first decade of the 21st century Nevatim saw some deployments by US Air Force units from Europe and the US in order to undertake joint exercises with the Israeli Air Force. One of the largest deployments took place in May 2006 and involved a USAF F-15 squadron from the Massachusetts Air National Guard and a USAFE F-16 squadron from Aviano AB, Italy. The Israeli side sent F-15Is from No. 69 'Hammers' Squadron, F-16Cs from No. 110 'Knights of the North' Squadron and the Nevatim-resident F-16A/Bs from the 'Defenders of the South' and 'Golden Eagle' Squadrons.

No. 103 'Flying Elephant' and No. 131 'Yellow Bird' Squadron (The Karnaf Wing)

The IASF's C-130 wing consists of two squadrons that share the fleet of C-130 Hercules. The squadrons are No. 103 'Flying Elephant' and No. 131 'Yellow Bird' Squadrons, both relocated in August 2008 from Lod to Nevatim in the Negev.

No. 103 'Flying Elephant' Squadron was initially formed at Ramat David in July 1948, during Israel's War of Independence, as a bomber and transport squadron. The squadron operated three Dakotas and one Douglas DC-5. A few days after its inauguration the squadron received a single Mosquito PR.Mk 33. In August 1948 the squadron received four Bristol Beaufighters smuggled from England during the 'shooting' of a fake movie about Australian Beaufighter operations in World War II. The squadron participated in the transport of supplies and personnel to Israeli towns under siege in the Negev Desert, as well as some bombing missions including one over Damascus,

The starboard side of C-130E serial number 208 with the 'Yellow Bird' Squadron insignia on the tail.
(Ofer Zidon)

Chapter 8

Unmarked C-130H serial number 436 takes off from Hatzerim. Note the refuelling probe extended from the port side above the cockpit, the Rafael TopLite FLIR turret below the nose and the refuelling pods under the outer wing stations.
(Ofer Zidon)

C-130H serial number 314 on the Nevatim apron. The aircraft displays the 'Flying Elephant' insignia on the left side of the fin, and elements of its electronic warfare suite on the rear fuselage.
(Ofer Zidon)

115

Modern Israeli Air Power

Paratroopers jump from a C-130H. Like many of the Karnaf aircraft, serial number 428 does not carry any identification markings beside its serial number on the tail and under the cockpit.
(Ofer Zidon)

Serial number 661, the first of four C-130J-30s ordered by Israel, displays its overall grey colour scheme, in contrast to the usual green, tan and sand over grey scheme of the current C-130H fleet. Delivery of the first IASF C-130J is due in spring 2014.
(Ofer Zidon)

executed with a Dakota. On 5 May 1949 the squadron was reinforced with C-46 transports and by the end of the month it had been relocated to Tel Nof. The squadron's missions comprised bombardment, transport and parachuting of supplies and troops, patrol, maritime patrol and anti-shipping operations. During the 1950s the squadron occasionally flew the B-17G and a Consolidated PBY Catalina flying boat.

In 1951 the squadron began to fly its Dakotas abroad to train the aircrews in long-range navigation. The aircraft participating in those sorties were painted in the colours of EL AL, Israel's national airline.

With the ageing of the Dakota fleet the IDF/AF began looking for a replacement. It was found in late 1955 in the form of the French-made Noratlas. The IDF/AF purchased three Noratlas that joined the 16 Dakotas of the 'Flying Elephant' Squadron during November 1955 and January 1956. The opening move of the 1956 Sinai Campaign was made on 29 October by a formation of the squadron's Dakotas that deployed elements of the paratroopers regiment in the Mitla Pass in Sinai. On 2 November the squadron parachuted a force of 100 paratroopers to take control of A-Tur airfield. After it was taken the squadron used this as a front-line airfield to support and supply IDF elements in the area of Sharm el-Sheikh.

During 1957 the squadron introduced a helicopter flight with two Hiller 360s, one Alouette II and two S-55s. With the success of the paratrooper operations a decision was made to enlarge the squadron's fleet and by the end of 1963 the unit operated no fewer than 24 Noratlas aircraft. In the weeks prior to the June 1967 War the 'Flying Elephants' flew maritime patrol missions. During the conflict the squadron participated in the transport of supplies and personnel and performed evacuation missions. On 7 June and for some nights after, three Noratlas aircraft participated in a search for three pilots that ejected over H-3 airfield in Iraq, without success.

During the War of Attrition the squadron flew transport missions bringing troops and supplies to Sinai, evacuations of wounded soldiers on the Suez Canal front, communications relay missions, illumination missions and more. During the October 1973 War the squadron was focused on search and rescue missions. The Noratlas flew patrols over the battlefield waiting to make contact with downed pilots.

On 15 August 1974 the squadron was relocated to Lod and joined all the other heavy transport squadrons of the IDF/AF, including its younger sister – No. 131 'Yellow Bird' Squadron. This latter was formed during the war to operate the 12 newly arrived C-130 and the two other C-130s received from the 'International' Squadron. On 3 July 1976 four C-130s from the 'Yellow Bird' Squadron participated in Operation Thunder Ball, the evacuation of Israeli hostages held at Entebbe Airport in Uganda. The aircraft flew special forces to the airfield and flew the released hostages back to Israel. The same year saw the retirement of the Noratlas and the conversion of the 'Flying Elephant' Squadron to the C-130. Both squadrons shared the C-130 fleet and each aircraft carried both squadron insignia on the tail. Since 1976 both squadrons have participated in countless operations, including the early 1980s airlift of Ethiopian Jews from airstrips in Sudan.

During the 1982 First Lebanon War the C-130 squadrons flew transport missions using front-line airstrips, electronic warfare missions, evacuation missions, aerial refuelling missions and others.

During the 2006 Second Lebanon War the C-130 performed low-altitude supply drops in action for the first time.

Modern Israeli Air Power

In the near future The 'Flying Elephant' Squadron will receive the new C-130J, while the 'Yellow Bird' Squadron will continue flying the older C-130H. In preparation for this development, by summer 2013 the 'Flying Elephant' Squadron has divested itself of all its remaining C-130s, these being passed on to the 'Yellow Bird' Squadron.

No. 116 'Defenders of the South' Squadron

This squadron was established in 1956 as No. 116 'Flying Wing' Squadron and became the 'Defenders of the South' Squadron on receipt of the F-16A/B at Nevatim in 2003. At first it was a combat unit based at Tel Nof. It flew P-51Ds, which were transferred from No. 101 'First Fighter' Squadron when it received the Mystère IVA. At that time the squadron's pilots were all instructors and commanders from the Flight School, flying in the squadron once a week to preserve their combat qualifications. In the months

F-16A serial number 107 ready for a training sortie. Serial number 107 is the highest-scoring Israeli F-16, with six and half Syrian MiG kills claimed during the 1982 First Lebanon War.
(Ofer Zidon)

prior to the Sinai Campaign the squadron enhanced its training regime and flew a number of operational sorties against Jordanian targets. With the outbreak of war on 29 October 1956 the squadron's Mustangs were the first to cross the border into Egypt. Their mission was the cutting of telephone wires connecting Sinai and Cairo. This was done using a weight connected by a steel cable to the P-51's tail. During the mission the cables were torn and the Mustangs completed the job using their propellers and wings. The squadron flew eight wire-cutting and 58 close air support sorties. The low-altitude ground attacks took a high toll – six of the Mustangs were shot down by small arms fire with four pilots rescued, one killed, and one captured to become a PoW.

On 15 January 1961 the P-51 era ended in the squadron and in the IDF/AF. The last propeller-driven fighter squadron in the ID/AF retired its Mustangs and converted to the Mystère IVA. For the second time, the squadron received aircraft from the 'First Fighter' Squadron, which were handed over to make way for the arrival of the new Mirage IIIs. During the June 1967 War the squadron flew 269 operational sorties including 49 air base attacks as part of Operation Moked on the first day of the war. One Jordanian Hunter was claimed as shot down and five Mystères were lost – according to Israeli accounts, three to AA fire and two to enemy fighters, with two pilots killed, two ejected and rescued, and one becoming a PoW in Egypt.

On 8 February 1968 the amalgamation process of the Mystère force was concluded and No. 116 'Flying Wing' Squadron remained the only Mystère IVA operator in the IDF/AF. It received the Mystères of the 'Valley' Squadron at Ramat David, which in turn became the first A-4 squadron. During the War of Attrition the squadron's Mystères flew 1,320 operational sorties, mostly ground attacks against front-line Egyptian targets including armoured vehicles, strongholds and artillery batteries. Three aircraft were lost, one of which in a training accident. On 18 July 1971 the Mystère IVAs were officially retired from Israeli service. By that time No. 116 'Flying Wing' Squadron was midway through a conversion process to the A-4E Skyhawk. The first flight of an A-4 in 'Flying Wing' colours took place on 1 April 1971.

The October 1973 War found the squadron in the process of conversion from the A-4E to the more advanced A-4N. During the war the squadron flew mixed formations of A-4E and N aircraft and carried out 823 operational sorties. 674 of these were high-risk ground attacks against Egyptian AA batteries and SAM sites. The squadron lost five Skyhawks, all on 9 October, with the squadron commander and another pilot killed, and two pilots captured to become PoW – one in Egypt and the other in Syria – while one pilot ejected and was rescued.

During the 1982 First Lebanon War the squadron flew air-to-ground missions and suffered no casualties. The peace agreement between Israel and Egypt forced the withdrawal of Israel from Sinai, including IDF/AF air bases. Three replacement bases were built in the south of Israel: Ovda, Ramon and Nevatim. The IDF/AF High Command decided that the 'Flying Wing' Squadron would be relocated from Tel Nof to become the first active squadron at Nevatim. The new base was inaugurated on 3 October 1983 with the first landing of one of the squadron's Skyhawks. A year, later No. 115 'Flying Dragon' Squadron also relocated from Tel Nof to Nevatim.

The next major event in the history of the squadron was assuming responsibility for the operational training course. The important mission of teaching new graduates how to fight was assigned to the squadron on 17 May 1985, with the arrival of the A-4s from No. 140 'Golden Eagle' Squadron that transitioned to the F-16A/B.

Modern Israeli Air Power

F-16B tail number 010 from No. 140 'Golden Eagle' Squadron takes off from Nevatim. As of August 2013, aircraft from this squadron have been taken on strength by No. 116 Squadron, to form a single pooled unit at Nevatim.
(Ofer Zidon)

By the end of 2002 the squadron had retired its Skyhawks and transferred them to No. 102 'Flying Tigers' Squadron at Hatzerim. In March 2003 the squadron received F-16A/Bs from No. 253 'Negev' Squadron, which in turn became Israel's first F-16I squadron. Now known as the 'Defenders of the South', No. 116 Squadron retained its operational training course after the receipt of the new fighter. In the last decade the squadron has flown thousands of operational sorties in all various conflicts in Gaza and Lebanon.

In 2013, No. 116 Squadron was merged with No. 140 'Golden Eagle' Squadron, a unit with an illustrious history dating back to 1950. At first the 'Golden Eagle' Squadron flew the North American T-6 Harvard and provided advanced training. The squadron also functioned as an emergency unit tasked to perform operational missions during wartime. In the 1956 Sinai Campaign the 'Golden Eagle' Squadron flew 46 operational sorties, mostly close air support. It lost three Harvards and the squadron commander was killed. The squadron's poor performance during the war was mainly a result of the incompatibility of its slow and vulnerable aircraft to the close support mission, and the manning of the squadron with under-trained air cadets and reserve pilots. As the IDF/AF became an all-jet force, the 'Golden Eagle' Squadron was closed down on 15 May 1959.

During the early 1970s the IDF/AF experienced huge expansion from nine to 14 fighter squadrons, with the arrival of new and advanced fighters in the form of the A-4 and F-4. This growth demanded an equivalent expansion of IDF/AF infrastructure and the training of an increasing number of pilots.

On 1 July 1973 the 'Golden Eagle' Squadron was reborn. In its new incarnation the squadron flew the A-4E Skyhawk from Etzion in Sinai, but retained its old title as the Operational Training Unit of the IDF/AF. The squadron was not active during the October 1973 War when its pilots and aircraft were dispersed among other A-4 squadrons at Ramat David and Tel Nof. The squadron's first operational training course began

Chapter 8

on 18 November 1973. On 14 April 1974 the 'Golden Eagle' Squadron executed its first operational sortie against Syrian targets around the Mount Hermon area.

In May 1976 the squadron replaced its veteran A-4E with the more advanced A-4N. The aftermath of the peace treaty between Israel and Egypt included the withdrawal from IDF/AF bases in Sinai and the construction of new bases in southern Israel. The squadron's aircraft and their crews were relocated from Etzion to Ramon on 5 November 1981. During the 1982 First Lebanon War the 'Golden Eagle' Skyhawks flew 224 operational sorties, comprising 179 close air support sorties and air strikes, eight sorties during the attack on Syrian SAMs in Lebanon, and some electronic warfare missions.

The third incarnation of the 'Golden Eagle' Squadron began on 3 August 1986. During the summer of 1985 the squadron was temporarily deactivated, as it prepared to receive the F-16A/B assigned from the Ramat David squadrons, in line with the arrival of the new F-16C/D at Ramat David. The operational training mission was assigned to No. 116 'Flying Wing' Squadron at Nevatim. The F-16-equipped squadron experienced a series of accidents in its early days on the type, with two pilots killed. On 10 October 1987 the squadron flew its first operational mission over Lebanon.

During the second half of the 1980s the IDF/AF underwent a transformation process when the balance shifted from the veteran delta fighters (Mirage, Nesher and Kfir), A-4 and F-4 to the new and more advanced F-15 and F-16 fighters. That process led

F-16A serial number 129 from No. 116 'Defenders of the South' Squadron lands at Nevatim. (Ofer Zidon)

to the decision to add an advanced operational training course employing a new and advanced platform. New graduates would fly the operational training course on the A-4 and than complete the advanced operational training course on the F-16, before moving to their assigned operational squadrons. The 'Golden Eagle' Squadron was selected to run the new course at Ramon and the first course began on 30 December 1990. Another step was taken in 1999 when the operational training course platform was changed from the A-4 to the F-16 Netz. It was the first time that new graduates of the flight course would experience the advanced F-16 as part of their training.

During the 1990s the squadron took part in operational activities over Lebanon and Gaza. It also organised and participated in multinational exercises with the US Navy, US Air Force, Turkish Air Force and Italian Air Force.

In April 2003, as part of preparations for the arrival of the F-16I at Ramon, the 'Golden Eagle' Squadron was relocated to Nevatim, to join No. 116 'Defenders of the South' Squadron – the second F-16A/B squadron, already stationed at the base. Both squadrons were thereafter responsible for operational and advanced operational training courses for new graduates. The 'Golden Eagle' Squadron was also responsible for the F-16A/B simulator that was constructed at the base by Elbit Systems. During the 2006 Second Lebanon War the squadron flew 1,000 operational sorties against Hezbollah targets in south Lebanon. During Operation Cast Lead in 2008 the squadron attacked targets in Gaza, including rocket launch sites, weapons smuggling tunnels and weapons storage sites.

While the aircraft and missions of No. 140 Squadron were transferred to No. 116 Squadron at the same base in August 2013, forming a single pool of F-16A/Bs, the 'Golden Eagle' traditions are expected to be revived in the future. As such, the first IASF F-35I unit is likely to be established as No. 140 Squadron in 2016. Nevatim has been confirmed as the first operating location for the F-35. Once both the F-35 and M-346 are in service it is expected that the F-16A/B will be retired altogether.

No. 120 'Desert Giants' Squadron

The roots of No. 120 'Desert Giants' Squadron lie in a flight that was established on 15 May 1963 in Lod, equipped with five Boeing 377 Stratocruisers. The aircraft were purchased from Pan American and were overhauled in IAI facilities. The flight, subordinate to Tel Nof, also received two Dakotas from No. 103 'Flying Elephant' Squadron. The new transport had three times the payload capacity of the Noratlas and possessed much longer range. The missions of the squadron were defined as transport and airlift, patrol and special missions.

On 1 July 1964 the flight became No. 120 'International' Squadron. The chosen name reflected its main role at the time – transport flights abroad. The Stratocruisers, known locally as Anak (giant) transported cargo and passengers between Israel, Europe, Iran and Africa. Between March and May 1964 one of the squadron's Stratocruisers flew 14 airlift sorties, dropping ammunition and medical supplies in Yemen, to help a revolt against the Egyptian-backed regime in the country. The Stratocruiser flew 14 hours to Yemen and back without being detected.

In early 1967 the IDF/AF purchased four surplus USAF Boeing C-97 Stratofreighters (the military version of the Stratocruiser), which entered service a year later after

Chapter 8

Boeing 707-300 serial number 275 flies in formation with three F-15 Baz fighters from No. 133 'Twin Tail' Squadron during a display for a graduation ceremony of new pilots. The tanker wears the old white-over-grey colour scheme. (Ofer Zidon)

Boeing 707-300 serial number 248 under maintenance at Nevatim. The tanker still wears the old white-over-grey paint scheme.
(Ofer Zidon)

Modern Israeli Air Power

IAI Seascan serial number 931 from the 'Desert Giants' Squadron's maritime operations flight. Radar is installed in the nose and trained Navy officers operate the on-board surveillance systems. (Ofer Zidon)

being overhauled in IAI facilities. Between 1967 and 1973 the IDF/AF also purchased another batch of five Boeing KC-97 Stratotankers, the aerial refuelling version of the C-97.

During the June 1967 War the squadron impressed a number of French Air Force Douglas C-47s that were undergoing maintenance, these being extracted from IAI lines. Most of the squadron's missions involved transport of ammunition and spare parts from France. The Stratocruisers also dropped water and fuel supplies to forces fighting in Sinai and flew communications relay missions.

In 1971 the 'International' Squadron received a pair of C-130 Hercules, which were flown alongside the veteran C-97. On 17 September 1971 an SA-2 missile shot one of the squadron's Stratocruisers down during a reconnaissance mission over Sinai. Seven crewmembers were killed and only one was able to bail out and survive.

The October 1973 War saw the squadron operate both the C-97 and C-130 in combat. Their primary missions were transport, evacuation and electronic warfare over the Egyptian and Syrian fronts. During the war the squadron received two Boeing 707s 'recruited' from IAI maintenance facilities, in order to help in the transport of supplies

Seascan serial number 929 is seen at its home base of Nevatim. Some of the various modifications to the original Westwind I business jet are clearly evident. (Ofer Zidon)

Chapter 8

Boeing 707-300 serial number 264 during aerial refuelling of a pair of F-16Ds from No. 109 'Valley' Squadron. The 707s wears the current overall dark grey colour scheme. (Yuval K.)

and equipment from abroad and to fly soldiers to and from the Sinai front lines. On 14 October 1973, No. 131 'Yellow Bird' Squadron was established, following the arrival of 12 C-130s from the US. The 'International' Squadron transferred its two C-130s to the new squadron.

The experience with the Boeing 707 during the 1973 war led to a decision to purchase more of these aircraft. By 1974 the 'International' Squadron was flying five Boeings on various missions. Their high capacity, high speed, and long range and endurance made them ideal for a wide range of activities including transport, electronic warfare, communications relay, command and control, aerial refuelling and more. On 4 July 1976 two of the squadron's 707s participated in the Entebbe raid, the rescue operation of Israeli hostages. One of the aircraft was equipped as an airborne hospital and the other was used for command and control.

In March 1975 a new front against Israel was opened when terrorists infiltrated from the sea and attacked Hotel Savoy on the Tel Aviv seashore, killing eight civilians and two soldiers. The IDF's response was to stage maritime patrols flown by IDF/AF aircraft with Navy officers on board to operate radars and other mission-specific equipment. In June 1977 a maritime operations flight was established within the 'International' Squadron. In 1978 this flight was equipped with the IAI Seascan, equipped with radar and other detection equipment. In the same year the Stratocruiser was retired from IDF/AF service and since then the squadron has been based around the Boeing 707. Through the years the 707 has taken part in countless operations in the aerial refuelling and command and control roles. On 1 October 1986 eight IDF/AF F-15s attacked PLO headquarters in Tunisia. It was the longest-range operation staged by the IDF/AF and Boeing 707s provided aerial refuelling and served as a forward command and control post.

Modern Israeli Air Power

The squadron has also taken part in many humanitarian aid missions, including the response to the major earthquake in Mexico in September 1985, the earthquake in Armenia in December 1988, and others. On 24-25 May 1991 the squadron, together with EL AL, operated a large-scale airlift of Ethiopian Jews. Around 15,000 people were brought to Israel using the squadron's Boeing 707s, together with C-130s and civilian EL AL airliners. Six of the squadron's Boeing 707s flew a dozen sorties, carrying almost 6,000 people.

The squadron flew six sorties to provide supplies of food and medical equipment to Rwanda following the Hutu-Tutsi war. Following the civil war in Yugoslavia, the government of Israel decided to provide aid to Muslim refugees in Bosnia, delivering 20 tons of medical equipment in a joint operation with the Royal Jordanian Air Force that commenced on 25 July 1995.

With the growing threat from Iran and other 'third-circle' nations, the IASF is increasingly prepared for long-range operations, including the conversion of some of the 'International' Squadron's passenger-configured 707s for the aerial refuelling role. In 2008 the squadron was relocated to Nevatim, together with all IASF heavy transport assets. Alongside its heavy transports, No. 120 'Desert Giants' Squadrons continues to operate the IAI Seascan in the maritime surveillance role, now reduced to a fleet of just three aircraft.

No. 122 'Nachshon' Squadron

No. 122 'Nachshon' (pioneer) Squadron was formed on 1 June 1971 at Lod. It assumed responsibility for Dakota operations from the 'International' Squadron and was initially known as the 'Dakota' Squadron. The squadron's designated missions were transportation, evacuation, and electronic warfare and special operations.

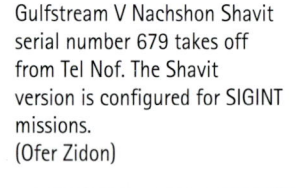

Gulfstream V Nachshon Shavit serial number 679 takes off from Tel Nof. The Shavit version is configured for SIGINT missions.
(Ofer Zidon)

Chapter 8

Gulfstream V Nachshon Shavit serial number 684. Note the lack of any insignia on the aircraft. (Ofer Zidon)

During the October 1973 War the squadron was reinforced with a number of Arava utility transports and flew hundreds of missions, including many electronic warfare missions over the Egyptian and Syrian fronts in order to jam the enemy's wireless communications. Since June 1974 the squadron has been responsible for the operational training course for transport pilots graduates. In August 1974 the squadron received two OV-1Ds, which were used for intelligence gathering and early warning operations, using their side-looking airborne radar (SLAR), optical camera and infrared sensor. The Mohawks flew with the squadron until 1982 when one was burnt out in a ground accident, and both were retired and returned to the US.

Gulfstream 550 Nachshon Eitam serial number 537 takes off from Ramat David. (Ofer Zidon)

Gulfstream 550 Nachshon Eitam serial number 569 takes off from Ovda during a joint Italian-Israeli exercise in January 2013. (Ofer Zidon)

The 1982 First Lebanon War saw the 'Dakota' Squadron flying its usual mix of transport, evacuation, electronic warfare and communications relay missions. A year later, on 30 October 1983, the squadron received its first batch of Arava 201s. The light transport played an important role throughout the squadron's operations. Between 1987 and 1989 a second batch of Arava 202s was accepted into the squadron. The 'Dakota' Squadron flew missions in all the conflicts during the 1990s, including Operation Accountability in 1993 and Grapes of Wrath in 1996.

On 3 August 2000 the Dakota was retired after over 50 years of service with the Israeli Air Force. A few years later, with the arrival of the King Air 200, the Aravas were also retired and the squadron was disbanded. However, on 26 June 2005 it was recommissioned to receive the new Gulfstream V and 550 airborne early warning, command and control and SIGINT aircraft. The name of the unit was changed to 'Nachshon' Squadron, but it retained its previous insignia. On 8 February 2009 the squadron was the last to be relocated to Nevatim and Lod was officially closed, the military area being returned to Ben Gurion Airport.

Chapter 9

OVDA AIR BASE

Brief history

Ovda is one of the fruits of the peace agreement between Israel and Egypt. Under the terms of the agreement, Israel withdrew from the Sinai Peninsula, leaving behind two of the more important IDF/AF air bases – Etzion and Eitam. Those bases were replaced by new bases constructed with US financial support and by US engineers in the south of Israel – Ovda and Ramon. Ovda is located 70km (43.5 miles) north of Eilat. The base was activated on 30 March 1982 and received two of the three squadrons that operated from Etzion in Sinai – No. 144 'Phoenix' and No. 149 'Smashing Parrot' Squadron – both flying the Kfir. In 1988, due to extensive budget cuts, Ovda's operational squadrons were transferred to Hatzor and Ovda became home of the Air Force School of Aviation Professions and the Officers School. In 2005 the base regained some of its former glory with the reactivation of No. 115 'Flying Dragon' Squadron, unofficially known as the 'Red Squadron', and the related Advanced Training Center that is run by the squadron. In the last decade all operational IASF squadrons have deployed many times to Ovda to fly against the 'Red Squadron'. The ATC also hosts deployments by foreign air forces including those of Greece, Italy, Poland and others.

No. 115 'Flying Dragon' Squadron

No. 115 'Flying Dragon' Squadron was formed in 1954, flying reconnaissance missions with Mosquitoes from Tel Nof. Prior to the Sinai Campaign the squadron flew a large number of photo-reconnaissance missions over Egyptian forces in the Sinai Peninsula, the Suez Canal and the Nile Valley area. After the 1956 Sinai Campaign the volume of the squadron's operational sorties decreased dramatically, while new IDF/AF jet fighters such as the Meteor and Mystère assumed responsibility for the photography mission, offering reduced risk and improved efficiency. As a result, the squadron was deactivated in 1958.

On 30 March 1969 the 'Flying Dragon' Squadron was reactivated, operating the newly purchased A-4. The War of Attrition demanded that the squadron become operational very rapidly and only a month after its activation the squadron participated in a ground-attack sortie against an Egyptian radar station. By the end of the war in August 1970, the squadron had flown 1,000 operational sorties.

Modern Israeli Air Power

F-16A serial number 234 from No. 115 'Flying Dragon' Squadron. The squadron runs the Advanced Training Center and is unofficially known as the 'Red Squadron' on account of its adversary role.
(Ofer Zidon)

During the October 1973 War the 'Flying Dragon' Squadron flew 750 sorties, mostly against the Egyptian SAMs located near the Suez Canal. The squadron lost seven aircraft, with five pilots killed and two captured by Egyptian forces to become PoWs.

In October 1984 the squadron was transferred to its new home base at Nevatim where it operated until it was decommissioned for a second time in 21 July 1994. Before its decommissioning the 'Flying Dragon' Squadron took part in Operation Accountability in July 1993, flying 38 day and night sorties.

The squadron's present incarnation began in February 2005 when it was reactivated as an F-16A/B unit, responsible for flying operations and the Advanced Training Center at Ovda. The unit's unofficial name is the 'Red Squadron' and it is unique as an IASF squadron in that it operates both fixed- and rotary-wing aircraft.

F-16A serial numbers 296 and 220 prepare to take-off from Ovda.
(Ofer Zidon)

Chapter 9

Advanced Training Center

In order to enhance training capabilities, by creating more professional, complex and realistic scenarios, it was decided to create a dedicated adversary unit that would operate as an Advanced Training Center for all IASF combat squadrons. The ATC together with No. 115 'Flying Dragon' Squadron was activated at Ovda in 2005. As well as its own flying duties, No. 115 Squadron operates the ATC, with training at the centre conducted under the control of the squadron, which also designs its syllabus. No. 115 Squadron provides the aggressor aircraft employed by the ATC.

The roots of a unit specialised in enemy tactics lie in the participation of Israeli Air Force units in large-scale international exercises such as the US Red Flag and the Canadian Maple Flag, during the 2000s.

Until the activation of the ATC, Red Force missions – namely the simulation of enemy interceptors, ground-attack forces and AA units – were carried out by regular Air Force squadrons. The quality of training was not satisfactory because the randomly chosen Red Force squadrons tended to fly using well-known Israeli tactics rather than genuine enemy tactics, so the need for a unit specialised in enemy tactics became obvious.

A Polish Air Force F-16C over Ovda. Serial number 4065 is on strength with the 10. Eskadra Lotnictwa Taktycznego at Lask. (Ofer Zidon)

Modern Israeli Air Power

In order to provide the IASF with full 'enemy simulation' the No. 115 Squadron structure is unique within the air arm, being the only squadron with a fixed-wing flight of F-16As and a rotary-wing flight of AH-1s, both flying in enemy style and practicing enemy tactics. These operate alongside enemy-style command and control and AA units. The squadron deploys AA batteries that simulate missile defences and uses Israeli ground forces to simulate guerrilla and terrorist threats. The ATC's training areas are in close vicinity to the base. They incorporate a high-altitude supersonic training area, a live-fire training area, an obstacle-free training area (especially important for low-flying aircraft and helicopter training), all within a few minutes' flight from the base.

The ATC has extensive experience in planning and coordinating exercises for Israeli Air Force units since 2005. It organises exercises of different sizes and complexities and includes mission planning and the use of No. 115 Squadron assets in either the Red (hostile) or Blue (friendly) Forces.

Training missions include one or more of the following components:
- Air-to-air training for aircraft and helicopters
- Air-to-ground training; navigation and use of GP bombs and smart weapons (GPS and electro-optical), including the use of realistic (inflatable) target vehicles
- Dust and low-visibility training
- Training with ground forces; search and rescue missions, AA units and counter-insurgency warfare including fighting in urban areas
- Thanks to the close proximity of the training area it is possible to integrate different missions within the same training period. The above features can be combined to create a very intensive and effective training package, which maximises the benefits from both the time and money invested in the deployment.

In December 2011, Tornado ECRs of the Italian Air Force's 50° Stormo were joined by Eurofighter Typhoons of 4° and 36° Stormo for a small-scale deployment to Ovda as part of Exercise Desert Dusk. MM7275 (36-11) is a Typhoon on strength with 10° Gruppo of 36° Stormo. (Ofer Zidon)

Chapter 9

Tornado ECRs from the Italian Air Force's 155° Gruppo, normally based at Piacenza, on the flight line at Ovda. (Ofer Zidon)

Two Italian Air Force AMX fighter-bombers at Ovda during their deployment to the base in March 2013. (Ofer Zidon)

133

PALMACHIM AIR BASE

Brief history

Palmachim was established on the coast of the Mediterranean south of Tel Aviv, in 1963. Due to its location the base is considered a 'space port' and hosts launches of missiles and satellites, pointed to the west, over the Mediterranean. Palmachim area is a known missile test range (and artillery test range) and over the years it has seen the launch of the Ofek series of satellites and trials of missiles including the Arrow anti-ballistic missile system. The unit understood to be associated with missile test work at Palmachim is No. 151 Missile Testing Squadron.

With the 1979 transfer of AH-1s from the recently established No. 160 'First Attack Helicopter' Squadron from Tel Nof, and the 1981 transfer of the Bell 205/212 utility helicopters of No. 124 'Rolling Sword' Squadron also from Tel Nof, Palmachim became the main helicopter base of the IDF/AF.

In 1985 a second AH-1 unit, No. 161 'Southern Cobra' Squadron, was activated at Palmachim alongside with the 'First Attack Helicopter' Squadron. In August 1994 the 'Rolling Sword' Squadron converted to the UH-60 and retired its Bell 212s. With the expansion of the advanced AH-64 attack helicopter force at Ramon, the two AH-1 squadrons were amalgamated in 2005 and No. 161 Squadron was shut down.

The Israeli Air Force UAV force has also been based at Palmachim since its early days. No. 200 'First UAV' Squadron was established in 1971 operating the Teledyne Ryan Model 124I Firebee drone (known locally as Mabat – glance) and the Northrop MQM-74 Chukar target drone (Telem – furrow). Since 1979 the UAV fleet has been based around Israeli-made UAVs ranging from the half-ton class IAI Scout (Zahavan – oriole) and Searcher (Hugla – partridge) and the Elbit Hermes 450 (Zik – spark), to the one-ton class IAI Heron. The four-ton class, turboprop-powered Heron TP is meanwhile based separately at Tel Nof.

On 16 September 2010 the 'Yanshuf/Yasur Simulator' squadron was inaugurated at the base. The squadron is responsible for all simulator training for the IASF assault helicopter force.

The base has seen operational service in all of Israel's conflicts since the early 1980s. It is considered the busiest IASF base, with its squadrons flying more than half of the IASF's annual flying hours. During the 2006 Second Lebanon War, Palmachim-based squadrons flew around 60 per cent of the 40,000 flying hours recorded by the IASF during the conflict. Today, UAVs participate in most IDF activities, including real-time information gathering for the ground forces' tank and infantry brigades.

No. 124 'Rolling Sword' Squadron

On 3 November 1956, three days before the end of the Sinai Campaign, the IDF/AF received two S-55 transport helicopters. They were transferred to No. 100 'Flying Camel' Squadron, to join two Hiller 360 helicopters already flying within a separate flight. On 1 January 1958 the helicopter flight became No. 124 'Rolling Sword' Squadron, under the command of Maj Uri Yarom. At first it only flew a pair of S-55s and one Alouette II. During May and June 1958 three S-58s medium transport helicopters were added to the squadron. At that time, with more capable and more dependable helicopters, the squadron began to develop its cooperation with special forces.

During January 1961 three more Alouette IIs joined the squadron. Almost two years later, at the end of December 1962, the 'Rolling Sword' Squadron experienced considerable enlargement with the arrival of 24 Sikorsky H-34s (the military version of the S-58), part of Israel's first arms deal with West Germany. The arrival of so many helicopters in such a short time (24 December 1962 to 11 March 1963) was a turning point for the squadron, which more than tripled in size from the 11 helicopters in its possession until that time. The older S-55s were retired on 2 April 1963, being transferred to the IDF/AF technical school for training future helicopter mechanics.

The squadron's missions during the June 1967 War comprised four combat transport operations in which 1,200 soldiers of the Paratroopers Brigade and other infantry forces were transported to the front lines; evacuation of 650 wounded soldiers; and search and rescue operations to recover 13 ejected pilots, seven of them from behind enemy lines.

A line of UH-60Ls in Palmachim. (Ofer Zidon)

Chapter 10

During the War of Attrition the 'Rolling Sword' Squadron's cooperation with infantry and special forces materialised in an operational context. The squadron led the pursuit of terrorists attempting to penetrate into the West Bank from Jordan, and before long the Jordan Valley was the backdrop to intensive operations by helicopters armed with machine guns and carrying small infantry teams to catch the terrorists.

On 15 December 1967 the IDF/AF's light helicopter force transitioned from the H-34 to the powerful new Bell 205, when the first batch of the new helicopters arrived at the port of Ashdod on board a cargo ship. The veteran H-34s were retired in October 1968. Two months later the Bell 205 received its baptism of fire during Operation Tshura (reward), the raid on the international airport in Beirut during the night of 28 December 1968. Thirteen Arab nations' airliners were blown up by Israeli special forces in retaliation for the hijacking of an EL AL aircraft to Algeria in July of that year.

During the October 1973 War the squadron generated 1,332 flight hours. Most of the squadron's sorties focused on the evacuation of wounded soldiers, including 25 downed IDF/AF aircrews and four enemy pilots. The squadron also participated in the pursuit of Egyptian commandos attempting to infiltrate behind Israeli lines. The intensity of use to which the Bell 205 was subject eventually revealed the helicopter's main weakness – its single engine resulted in inferior performance at high altitudes and high temperatures. In light of this, the IDF/AF began to search for a Bell 205 replacement. This came in the form of the Bell 212, a twin-engine helicopter based on the fuselage of the Bell 205. As well as twin engines, the new helicopter's advantages included modern navigation systems that allowed operations in adverse weather conditions and over the sea. A contract for the purchase of around 60 helicopters was signed on March 1974 and deliveries followed between April 1975 and March 1976.

On 23 August 1981 the squadron moved from Tel Nof to Palmachim where it is still stationed today. During the 1982 First Lebanon War the 'Rolling Sword' Squadron, as usual in large-scale conflicts, focused on rescue and evacuation missions.

Following its acquisition of long-range Harpoon missiles with over-the-horizon targeting capability, the Navy required a shipborne helicopter that could operate beyond

UH-60L Yanshuf 3 serial number 571 waits to collect a group of infantry in the Negev Desert. (Ofer Zidon)

Modern Israeli Air Power

A demonstration of the UH-60's ability to carry slung loads is provided by serial number 575, one of the most recent Yanshuf 3 deliveries.
(Ofer Zidon)

UH-60L serial number 818. This airframe is part of the second batch of surplus US Army helicopters. It began its Israeli service as serial number 918 and in the Olive Drab scheme. After local modification it became 818 and was repainted in the desert camouflage scheme.
(Ofer Zidon)

Chapter 10

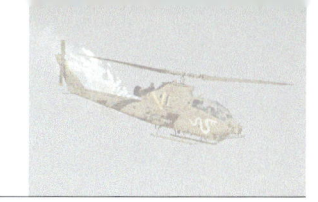

UH-60L Yanshuf 2 serial number 808 deploys special forces. Some of the earlier UH-60s retain their US Army Olive Drab colours.
(Ofer Zidon)

the warship's radar range. The Aérospatiale HH-65A Dolphin was chosen to fulfil the requirement. On 26 July 1985 the 'Rolling Sword' Squadron received a pair of HH-65As and formed a maritime operations flight within the squadron. In April 1987 the flight became a separate squadron – No. 193 'Defenders of the West' Squadron – and moved to Ramat David, closer to the Navy base in Haifa.

The squadron's most recent incarnation began in August 1994, with the arrival of surplus US Army UH-60As. The new helicopter is a much more powerful, stable and survivable platform than its predecessors. The squadron flew both UH-60s and Bell 212s for a few years, before the latter were retired on 15 July 1998. Some of the survivors were sold to foreign countries. In the last two decades the squadron has flown thousands of search and rescue and evacuation missions, coming to the aid of soldiers in wartime and civilians in distress in times of peace.

No. 161 'Southern Cobra' and No. 166 'Hermes' Squadron

Starting in 2003, Palmachim has seen numerous operations flown by the Elbit Hermes 450 UAV, operated by No. 161 'Southern Cobra' Squadron and No. 166 'Hermes' Squadron.

No. 161 Squadron previously served as the second IDF/AF AH-1 squadron, variously known as the 'Southern Cobra' or 'Fighting Family' Squadron, which operated the attack helicopters from Palmachim from 1985 until 2005, when it became a UAV operator.

No. 166 Squadron was established in 2003, and its introduction of the Hermes 450 permitted the IAI Scout to begin to be phased out of active-duty service.

Entering service with the IDF/AF in 1999, the Hermes 450 is a mid-size UAV with an endurance of less than 24 hours and operating altitude below 20,000ft (6,096m).

Modern Israeli Air Power

A Hermes 450 flies over Palmachim. Note the impressive tail art on the UAV, which is restricted to the aircraft in use with No. 166 'Hermes' Squadron.
(Ofer Zidon)

The Hermes can operate autonomously, tracking a pre-defined flight course. It can use interchangeable payloads including regular camera, infrared camera and synthetic aperture radar for adverse weather/night conditions. Communication between the UAV and the control unit can be conducted via satellite link. In May 2010 Elbit announced that it had won a contract to sell its new Hermes 900 to the IASF. The USD 50-million contract included provision of the new Hermes 900 as well as additional Hermes 450s over a three-year period. In December 2012 the IASF selected the Hermes 900 as its next-generation medium-altitude, long-endurance UAV, placing a second contract for an unconfirmed number of the vehicles and their support systems. As of summer 2013 the Hermes 900 was undergoing development trials at Palmachim.

A Hermes 450, operated by No. 161 Squadron, on display at Tel Nof.
(Ofer Zidon)

140

Chapter 10

Hermes 450 serial number 318 flies over the Mediterranean, displaying its upper-wing camouflage scheme.
(Ofer Zidon)

A line of No. 166 Squadron Hermes 450 UAVs at Palmachim.
(Ofer Zidon)

141

No. 200 'First UAV' Squadron

During the War of Attrition IDF/AF fighter squadrons fought hard to counter Egypt's deployment of SA-2 and SA-3 SAM systems. The high toll of aircraft losses led the IDF/AF to explore new solutions to meet its requirement for intelligence from the front lines. In July 1970, a month before the War of Attrition ended, a contract was signed with Teledyne Ryan to develop a new and improved long-range photo-reconnaissance UAV, based on the Firebee. The resulting Model 124I was Israel's first UAV and arrived a year later, in July 1971. With its arrival No. 200 'First UAV' Squadron was established on 1 August 1971 at Palmachim.

In September 1971 a first operational sortie was executed over the Suez Canal area and the squadron became officially operational in December 1971. Although two of the UAVs were shot down over Egypt, the system was considered a success and the IDF/AF then began to look for a decoy UAV to assist in operations against SAMs. The solution was found in the procurement of the MQM-74 Chukar, 27 of which were purchased.

During the October 1973 War the Chukars were successfully used on the northern and southern fronts. A total of 23 Chukar sorties were launched, and five were shot down. The Chukars were launched in groups of two or four UAVs and each could draw up to 20-25 firings of Egyptian SAMs, proving the effectiveness of the solution. The Firebees were also used extensively for photography missions, flying 19 reconnaissance sorties for the loss of 10 shot down. Nevertheless, the concept was again considered a success and the IDF/AF decided to purchase more Chukars and Firebees after the war.

The growing use of UAVs in real-time intelligence-gathering missions led the IAI to develop its own UAV with the revolutionary capability of broadcasting photos from a stabilised camera. The result was the IAI Scout, 20 examples of which were purchased and delivered to the squadron in June 1979. On 21 June 1981 the squadron's Scout flight was declared operational and from then on the Scout was responsible for most of the squadron's operations.

Heron serial number 227. This UAV is used for maritime operations, using a specialised radar that is located within the fairing under the fuselage. (Ofer Zidon)

Chapter 10

Heron serial number 267 takes off from Palmachim. Specially trained UAV take-off/landing operators conduct this procedure.
(Ofer Zidon)

Heron serial number 298 is towed to the runway at Palmachim.
(Ofer Zidon)

143

The Scout's first operational mission took place in 1981, flying into Lebanon and transmitting real-time imagery of the Syrian SAMs that had been transported to Lebanon. On 14 May 1981 a Firebee was flying a photographic mission over the Bekaa Valley. A Syrian MiG-21 was scrambled against the Firebee and during its attempts to hit the UAV it stalled and crashed, becoming the first jet fighter lost to a UAV in the Middle East.

During the 1982 First Lebanon War the Scouts were deployed to the north of Israel and participated in many missions, including collecting intelligence on SAM sites and other targets, damage assessment, and targeting AFVs. The Firebees flew only three photo-reconnaissance sorties but two were shot down, reflecting their increasing vulnerability, even to the relatively old weaponry in the Syrian arsenal.

On 16 July 1992 the squadron received the next generation of indigenous UAVs in the form of the IAI Searcher. This type was flown alongside the Scouts. Operation Accountability in 1993 saw 27 operational sorties by the squadron's UAVs. The continued success of the UAV operations led to the implementation of new capabilities including night-time and colour photography. Another development was the arrival of the long-range Searcher II in July 1999.

At the beginning of the 2000s the levels of violence in the West Bank and Gaza escalated, and with it the demand for UAV operations was increased too. UAVs became a crucial part of every IDF/AF operation against terror targets, acquiring and tracking targets and completing damage assessment after the target was attacked. On 16 August 2005 the first IAI Heron, an advanced new one-ton class UAV, was flown by the squadron. The official acceptance ceremony of the Heron took place on 7 March 2007. The Heron is still being developed and adapted to new missions including maritime patrol.

A further element of No. 200 'First UAV' Squadron is the UAV School, which has operated alongside it at Palmachim since April 2004. The school trains both domestic and foreign UAV operators, a task previously undertaken at Ein Shemer. In June 2011 it was decided to relocate this effort to Palmachim, where a new, purpose-built facility was completed. As well as training UAV operators, the school provides an officer training course for operators, conducts continuation training, trains UAV instructors and maintains its own UAVs and ground stations. Initially equipped with the Scout, the school switches to the Hermes and Heron on its arrival at Palmachim. The UAV operator training course lasts six months.

'Yanshuf/Yasur Simulator' Squadron

In the past, Israeli Air Force assault helicopter aircrews were training in emergency procedures using simulators based in the US. Beside the differences in cockpit layout, the Israeli crews were only able to practice once every 18 months. One of the results of the 4 February 1997 mid-air collision between two CH-53 Yasur 2000 helicopters, resulting in the death of 73 soldiers and crew, was the decision to purchase a modern assault helicopter simulator under a USD 35-million Foreign Military Funding contract. The simulator was designed to train both UH-60 Yanshuf and CH-53 Yasur aircrews, using two interchangeable cockpits. The simulator's realistic 3D simulation is created using hydraulic jacks and a 'real world' image projected on screens with a width of 230°.

Chapter 10

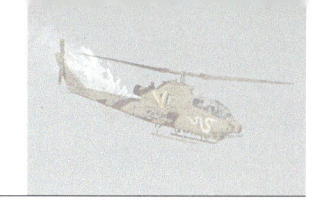

The cockpit of the UH-60 in the Yanshuf/Yasur simulator at Palmachim. The scenery is displayed on a 230° screen. (Ofer Zidon)

The simulator control station. (Ofer Zidon)

145

Modern Israeli Air Power

The simulator was installed in 2004 and training began in February 2005. The simulator provides 6,700 sorties per year and operates for between 11 and 17 hours each day. The simulator's main uses include:
- Basic competence and basic procedures
- Normal and emergency procedures
- Instrument flight rules/instrument meteorological conditions (non-visual) flight training
- Tactical mission training
- Mission rehearsal

Training routine consists of a training day once every three months. The training day comprises three sorties, and each pilot must fly at least six sorties each year to remain qualified. The simulator helps in training complex procedures such as in-flight refuelling, low-level flight, and high-voltage electric wire avoidance. Among the tactical mission training scenarios offered are combat search and rescue, formation flight, in-flight refuelling, landing in dust, and night vision goggles.

The visual display is based upon an updated geographical database with a high level of accuracy, allowing crews to rehearse a mission in the simulator before they take off in the real world. The crews can check flight routes in and out of the target area, look for emergency landing zones on the way, and become acquainted with the target and its vicinity. Training is conducted by dedicated trainers – usually female soldiers in regular service. They learn their trade for a year and must serve for two more years in the squadron as qualified teachers.

Currently, 15 per cent of an IASF helicopter pilot's flight hours must be flown in the simulator, and this will increase to 20 per cent in the future.

A sharp manoeuvre is executed by the simulator trainee. (Ofer Zidon)

Chapter 11

RAMAT DAVID AIR BASE

Brief history

The origins of Ramat David lie with the RAF. The RAF station, officially named after British Prime Minister Sir David Lloyd George, was also known as RAF Ramat David because of its proximity to the Jewish kibbutz of Ramat David.

The base was built in 1941 as a backup to Haifa, after this was attacked by German bombers. The new base was located near one of the stations on the Haifa–Damascus train line and over the Mosul–Haifa oil pipeline.

On 15 May 1948 (the day the British mandate over Palestine ended) the base was attacked by Egyptian Spitfires, mistakenly thinking the base was already in Israeli hands. The attack damaged a few buildings and small number of RAF Spitfires on their parking aprons. Four British soldiers were killed. A second Egyptian air attack was met, later that day, by British Spitfires shooting down five Egyptian Spitfires.

The British forces' retreat from Palestine had ended by 15 May 1948, with RAF Ramat David, because of its proximity to the main British harbour of Haifa, hosting fighters that covered the British withdrawal from Palestine. The base was handed over to the Israeli Air Force on 23 May 1948.

The first to move to Ramat David was the 'Tayeset Hagalil' (Galilee Squadron), which operated three Auster light aircraft over the Galilee area, helping Jewish towns fighting against Arab forces. After two weeks the squadron left the new base and returned to its original airfield in Yavniel.

On 2 July 1948 Ramat David saw the arrival of its first regular unit – Tayeset 103 (No. 103 Squadron, later known at the 'Flying Elephant' Squadron) – with three Dakotas modified for bombing operations. These aircraft flew their first missions on 8 July against Quneitra in the Golan Heights and on 16 July against Damascus with great success.

On 16 July No. 69 'Hammers' Squadron joined the base with three B-17G bombers. During the same month one Mosquito and four Beaufighter bombers joined Tayeset 103.

The War of Independence ended in March 1949. The IDF/AF underwent organisational changes that included the transfer of Ramat David's heavy squadrons to Ekron (Tel Nof). Hatzor was closed and No. 101 'First Fighter' Squadron was transferred to Ramat David on 30 May 1949.

With the accumulation of Spitfires and P-51Ds it was time to open a second fighter squadron. No. 105 'Scorpion' Squadron was established on 1 December 1950. The new unit was focused on the operational training of new graduates. With the arrival of another batch of Spitfires purchased in Italy and more P-51Ds purchased in Sweden, No. 101 'First Fighter' Squadron operated the new Mustangs and handed over its remaining Spitfires to a third squadron, No. 107 'Orange Tail' Squadron. The new squadron was activated in January 1953 and assumed the operational training role from the 'Scorpion' Squadron, which became a fully operational Spitfire squadron alongside the 'First Fighter' squadron with Mustangs.

A historical event took place on 7 June 1953 with the commissioning of No. 117 'First Jet' Squadron at Ramat David. The new squadron operated a mix of newly built Meteor variants: T.Mk 7 trainers, F.Mk 8 fighters and FR.Mk 9s for reconnaissance missions. The new squadron joined the three older fighter squadrons to make Ramat David the leading fighter base of the IDF/AF.

The purchase of French-made jet fighters – the Ouragan and Mystère IVA – in 1956 brought changes to the structure of the IDF/AF. No. 101 'First Fighter' Squadron was transferred to Hatzor to be equipped with the new Mystère IVA, leaving its Mustangs to the 'Scorpion' Squadron in Ramat David. No. 119 'Bat' Squadron was established in Ramat David, flying the newly arrived Meteor NF.Mk 13 night fighters.

A few months before the Sinai Campaign No. 110 'Knights of the North' Squadron was reactivated with its Mosquitoes, and No. 69 'Hammers' Squadron with its B-17Gs was deployed to Ramat David. Another addition to the base was two French Air Force Mystère IVA squadrons deployed to Ramat David during the Sinai Campaign in October 1965. Their mission was to help defend Israel's airspace from enemy attacks.

The Ramat David squadrons were involved in both the opening moves of the Sinai Campaign. A Meteor NF.Mk 13 from No. 119 'Bat' Squadron shot down an Egyptian Dakota carrying many high-ranking commanding officers of the Egyptian Army en route from a meeting in Syria, and Meteors from No. 117 'First Jet' Squadron flew top cover for 16 Dakotas that dropped elements of the Paratroopers Brigade over Sinai.

With the end of the Sinai Campaign came the end of the 'Hammers' and 'Knights of the North' Squadrons in their 1950s incarnations as B-17 and Mosquito operators. The 'Scorpion' Squadron was also closed down and its Mustangs retired. On the other hand, a new squadron was activated in December 1956 and joined the 'First Jet' and 'Bat' Squadrons that were already active at Ramat David. This was No. 109 'Valley' Squadron, which flew the Mystère IVA. 1957 saw the transfer of No. 119 'Bat' Squadron to Tel Nof and the reactivation of the 'Knights of the North' to fly the brand-new Vautour II bomber. Another change took place at Ramat David when the 'First Jet' Squadron was re-equipped with the new Mirage III. The 'Orange Tail' Squadron was reactivated to receive the old Meteors of the 'First Jet' Squadron and served as the jet OTU squadron until 1964, when it was closed down and the Ouragan-equipped No. 113 'Hornet' Squadron assumed its OTU mission at Hatzor. With the accumulation of Ouragans, the 'Orange Tail' Squadron was recommissioned in October 1965 to fly the French jet.

Ramat David's squadrons took part in Operation Moked, the attack on enemy air bases that opened the June 1967 War. 'Orange Tail' Ouragans, 'Valley' Mystères, 'Knights of the North' Vautours and 'First Jet' Mirage IIIs all flew ground-attack missions, with only four Mirages left at the base to fulfil the QRA mission. Since a major part of the enemy air forces had been destroyed, the rest of the war saw plenty of close air support missions to assist IDF ground forces. An exception was the attack on the Iraqi H-3 airfield conducted by four Vautours from the 'Knights of the North'

Squadron with top cover by a pair of 'First Jet' Squadron Mirage IIIs. Two Iraqi Hawker Hunters and two MiG-21s scrambled to intercept the attack and during the combat over the base the Israelis claimed one Hunter and one MiG-21 were shot down by a Mirage III and another Hunter shot down by a Vautour. (According to Iraqi documents, one Hunter crashed on take-off, another was damaged in air combat and one MiG-21 was claimed by two Israeli pilots – and 'confirmed' as shot down – but actually landed in a damaged condition.) The next day saw another attack on the same base with one Iraqi Hunter shot down for the loss of two Vautours and one Mirage III shot down by Iraqi fighters.

The June 1967 War signalled the end of the French era and the beginning of the American era in the IDF/AF. The first US-made jets – the A-4 Skyhawk – arrived in Israel by sea and were transferred from the port of Haifa to Ramat David where they were used to re-form No. 109 'Valley' Squadron as the first Israeli Skyhawk unit. The Ouragan-equipped 'Orange Tail' Squadron was deactivated once more and its aircraft were transferred to the 'Hornet' Squadron at Hatzor.

The second step of the American era was the reactivation of No. 69 'Hammers' Squadron as the IDF/AF's second F-4 operator in 23 October 1969. A month later, 'Hammers' pilots were the first Israeli crews to claim a MiG in the new fighter. In March 1971 the 'Knights of the North' gave up its veteran Vautours in favour of the A-4, to become the second Skyhawk squadron at Ramat David.

Ramat David entered the October 1973 War with four squadrons: 'Valley' and 'Knights of the North' equipped with A-4s, 'Hammers' flying the F-4, and the 'First Jet' Squadron still flying its vintage Mirage IIIs. Since it was the main base in the north of Israel, Ramat David was responsible for the northern front – the Lebanese and Syrian borders. The first mission was to stop Syrian armoured columns moving fast to the west of the Golan Heights and deep into Israeli territory, where there were only a few dozen Israeli tanks to fight them. The A-4 squadrons were used primarily in this mission while the F-4s were used for longer-range assignments such as the attack on the Syrian Army HQ in Damascus. The Mirages of the 'First Jet' Squadron found themselves intercepting enemy MiGs on the northern and southern fronts, with very good results. Ramat David units claimed 66 enemy MiGs shot down during the war, 11 by 'Hammers' Phantoms and 55 by 'First Jet' Mirages. However, the toll of the 1973 war on the base was very high. Over 20 A-4s were lost with 13 pilots killed, while nine F-4s were lost together with two crews killed and four captured.

During the war, emergency shipments of F-4Es from USAF units in Europe and A-4s from US Navy units in the Mediterranean arrived in Israel. The Phantoms were accepted at Ramat David and some of them took part in the final days of the fighting, flying with the 'Hammers' Squadron. The new Skyhawks were painted in Israeli colours and helped fill in for the A-4 losses suffered during the war.

On 28 July 1977 the 'Valley' Squadron received its first 11 Kfir C1s to become operational with the new fighters. Next, the 'First Jet' and 'Knights of the North' Squadrons received the new F-16A/B during 1980. Both squadrons took part in Operation Opera in 1981, the attack on the Iraqi nuclear reactor at Osirak. The two squadrons also participated in the 1982 First Lebanon war, claiming almost 40 Syrian MiGs destroyed, most of them after the attack against Syrian SAM positions in the Bekaa Valley in Lebanon on the second day of the war. Both squadrons transitioned to the more advanced F-16C in 1987.

No. 109 'Valley' Squadron was the last to enter the F-16 era. The Kfir outfit was deactivated in 1985 and activated again in 1991 to receive the two-seat F-16D. In 1999 the 'Valley' Squadron was the first to use the Rafael Litening navigation and targeting system.

In June 1991 the base's Phantom squadron – No. 69 'Hammers' Squadron – was relocated to Hatzerim to join the 'Orange Tail' Phantoms in the south of Israel.

During 1990 Ramat David saw the arrival of No. 190 'Magic Touch' Squadron. The squadron, which flew the Defender attack helicopter, required a base closer to its main theatre of operations in Lebanon, so a decision was taken to relocate it from Ramon. In August 1994, with the expansion of the AH-1 fleet, the Defenders were handed over to the Flight School and the squadron was closed down. However, a year later, in April 1995, the base saw the inauguration of another helicopter squadron, No. 193 'Defenders of the West' Squadron, dedicated to maritime operations. It initially flew two American-provided HH-65s provided by the US and in 1996 transitioned to the more advanced AS.565MA Panther helicopters. The squadron was based at Ramat David to make use of its proximity to Haifa harbour and the Navy Sa'ar corvettes that host the helicopters.

Currently, three F-16 squadrons operate from Ramat David: No. 109 'Valley' Squadron (F-16D), No. 110 'Knights of the North' Squadron and No. 117 'First Jet' Squadron (both F-16C), while No. 193 'Defenders of the West' Squadron continues to operate maritime helicopters.

All these squadrons played a major part during the last decade in conflicts fought against Hezbollah in Lebanon and against Hamas in Gaza.

No. 109 'Valley' Squadron

No. 109 'Valley' Squadron was established in June 1951 at Hatzor. In its first incarnation it flew the British-made Mosquito. The Mosquito's combination of lightweight wooden structure and twin engines provided speed and long range, making it ideal for photo-reconnaissance missions and long-range patrols. On 3 and 4 September 1953 the 'Valley' Squadron's Mosquitoes conducted their first long-range aerial photography flights over Egypt. The first sortie was flown over Alexandria and its port, while the second took place over Cairo and the surrounding airfields. 1956 saw the end of Israel's Mosquito era and a decision was made to re-equip the 'Valley' Squadron with the newly arrived Mystère IVA. The 'Valley' Squadron was relocated from its original Hatzor base and began its new incarnation as the IDF/AF's second Mystère IVA squadron at Ramat David. In April 1957 it was declared operational and entered the IDF/AF fighter squadrons' QRA rotation. During the first half of the 1960s the squadron developed new ground-attack techniques with the Mystère IVA, including night strikes, precision bombing, and accurate strafing, to become the IDF/AF's leading air-to-ground squadron. These techniques were put to the test in a number of border incidents against Syrian forces, and demonstrated great success.

During the June 1967 War the squadron took part in Operation Moked, attacking Egyptian and Syrian airfields including Damascus air base. Some of the squadron's aircraft were deployed to Tel Nof, closer to the Egyptian front, providing better access to distant targets in Egypt. The squadron finished the first day of the war having flown

59 operational sorties. Some aircraft were damaged and one was lost after being hit first by Syrian MiG-17s and then by Lebanese Hunters, with its pilot becoming a PoW in Lebanon. During the remainder of the war the squadron's sorties consisted mainly of close air support missions. The 'Valley' Squadron flew a total of 341 operational sorties in six days of war and lost two aircraft, with one pilot captured in Lebanon and one rescued.

The 'Valley' Squadron was selected to become the first in the IDF/AF to be equipped with the new A-4 Skyhawk. 29 December 1967 symbolised the transition of the IDF/AF from the French era to the American era, with the arrival of four newly built A-4Hs on board a cargo ship at Haifa harbour. The aircraft were transferred to Ramat David, assembled and flew their maiden flight with the 'Valley' Squadron the next day.

During the War of Attrition the Skyhawks were used primarily in the ground-attack role, attacking Egyptian SAM batteries and ground forces. The squadron flew a total of 2,077 operational sorties, some of them against Syrian targets. It claimed two Syrian MiG-17s shot down: one using cannon fire and the other with unguided rockets.

The next chapter in the squadron's history took place during the October 1973 War. During the 19 days of fighting the squadron flew 944 operational air-to-ground sorties, most of them high-risk missions against SAM batteries and well-protected Egyptian and Syrian targets. The result was a high number of causalities: 13 aircraft lost,

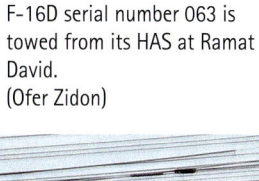

F–16D serial number 063 is towed from its HAS at Ramat David.
(Ofer Zidon)

Modern Israeli Air Power

with seven aircrew killed, five downed and rescued safely, and one captured by Syria, before being tortured and losing his left leg while in prison. Two of the squadron's pilots received the Medal of Distinguished Service for their exemplary bravery in the line of duty during the war.

After the October 1973 War the squadron took part in the daily fighting against Palestinian terrorists that took over the southern part of Lebanon and made the area a base for attacks against the northern cities of Israel.

On 20 July 1977 the 'Valley' Squadron transferred its last A-4 to its 'Knights of the North' sister squadron in order to make way for a new aircraft type, the Kfir. The 'Valley' Squadron received Kfir C1 airframes from No. 101 'First Fighter' Squadron at Hatzor, when the latter received the improved Kfir C2. On 28 July 1977 the first Kfir C1 landed at Ramat David. The Kfir was a natural addition to the squadron, following its Mystère IVA and A-4 predecessors in the air-to-ground attack and close air support roles. The squadron took part in numerous operational sorties during the 1970s and 1980s, including Operation Litani in 1978 and the First Lebanon War in 1982. In 1982 the squadron flew 342 operational sorties. With the arrival of the Kfir C2 and the Python 3 air-to-air missile, after many years the squadron received an air-to-air competence to add to its excellent air-to-ground skills. On 24 July 1985, after eight years of flying the Kfir, the squadron was closed down once again.

F-16D serial number 061 lands at Ramat David, armed with a pair of GBU-10 laser-guided bombs and AIM-9 AAMs. (Ofer Zidon)

Chapter 11

F-16D serial number 022 conducts an afterburner take-off from Ramat David.
(Ofer Zidon)

F-16D serial number 050 on its way to the end of the Ramat David runway, armed with a GBU-15 guided bomb and an AIM-9 Sidewinder AAM.
(Ofer Zidon)

153

Modern Israeli Air Power

F-16D serial number 088 from No. 109 'Valley' Squadron visited Hatzor for combined training with No. 105 'Scorpion' Squadron. Note the small practice bomb under the centreline pylon.
(Ofer Zidon)

The 'Valley' Squadron was reactivated in its present incarnation as an F-16D operator on 2 July 1991. Again it received its two-seat F-16s from No. 101 'First Fighter' Squadron. The new 'Valley' Squadron soon became one of the leading squadrons in the IDF/AF, and won first place in many internal competitions, including gunnery and precision bombing contests.

During the Operation Accountability July 1993 the 'Valley' Squadron flew 26 operational air-to-ground sorties. The two-seat F-16D was equipped for the launch of precision-guided munitions, and as a result the squadron was kept very busy during the operation, attacking enemy ground targets. In comparison, the two single-seat F-16C squadrons at Ramat David flew fewer operational sorties.

In the last decade the squadron has received new weapons including GPS-guided bombs, improved laser-guided munitions and others, to become the main attack squadron within Israel's sensitive northern border region. The squadron has taken a key role in all recent conflicts including the Second Lebanon War, Operation Cast Lead and others.

No. 110 'Knights of the North' Squadron

The first incarnation of No. 110 'Knights of the North' Squadron lasted between 2 August 1953 and 5 October 1955. The squadron received a number of Mosquito fighter-bombers that were flown from Hatzor.

The squadron served in the night fighter role and as an OTU for new graduates of the flight course. During this two-year period the young squadron suffered the loss of six pilots in accidents. Two were lost over the Mediterranean on a night navigation sortie and another two were killed in the same area while searching for the first crew.

On 5 October 1955 the squadron was deactivated and its OTU role was transferred to No. 109 'Valley' Squadron. The Mosquitoes were stored at IAI's facilities in Lod.

Chapter 11

F-16C serial number 377 from the 'Knights of the North' Squadron in the QRA shelter at Ramat David. The jet is in air-to-air configuration, with AIM-9 Sidewinder and Python 5 AAMs.
(Ofer Zidon)

F-16C serial number 383 taxis back to its HAS in pouring rain at Ramat David.
(Ofer Zidon)

155

Part of the IDF/AF's preparations for the Sinai Campaign in October 1956 was the reactivation of the 'Knights of the North' as a fighter-bomber squadron. Since the Mosquito was fast, possessed long range and could carry a large weapons load, it was used in the long-range fighter-bomber role. The 'Knights of the North' operated their Mosquitoes in close air support and ground-attack missions all over Sinai. They attacked Egyptian armour convoys and facilities in the south of the Sinai Peninsula near Sharm el-Sheikh. Despite their wartime success, when the war ended it was decided to retire the old Mosquitoes and re-equip the squadron with a new and twin-engine jet bomber, the French-made Vautour II.

The squadron's Vautour era began on 19 January 1958 at Ramat David, with actual flights starting three months later, on 7 April 1958. On 7 June 1959 a pair of 'Knights of the North' Vautours was involved in the squadron's first aerial battle, fought, according to Israeli sources, against four Egyptian MiG-17s. The MiGs were chased over the Israeli Negev, but escaped back to Egyptian territory before the Vautours could hit them. On January 1960 it was decided that the Vautour would assume the photo-reconnaissance role and the 'Knights of the North' thereafter visited every corner of the Middle East. On 23 January 1962 the first night photography sortie took place over Cairo. Another important date in the squadron's history was 16 March 1962, when a first operational bombing sortie took place against a Syrian stronghold near the Sea of Galilee.

During August 1963 the squadron received examples of the Vautour IIN from the 'Bat' Squadron, when the latter transitioned to the Mirage III. With this move, the 'Knights of the North' flew all three versions of the Vautour: the IIA tactical fighter, IIB bomber, and the IIN night fighter.

The squadron's long-range attack aircraft played a major role in the June 1967 War. They attacked air bases, radar stations and missile batteries in Egypt and Syria and flew close air support missions to assist IDF ground forces on the Syrian front. The squadron carried out long-range sorties against Rass Banes and Luxor air bases in Egypt and H-3 air base in Iraq. During the strike on Iraq on 6 June 1967 one of the squadron's Vautours claimed one Iraqi Hunter as shot down while it was defending the air base, although according to Iraqi documentation, no Iraqi Hunters were lost in combat during the engagement.

During the June 1967 War the squadron flew 278 operational sorties, with the loss of three Vautours. Two aircrew were killed, one ejected over H-3 air base in Iraq and was captured, and a fourth pilot ejected over Syria and was captured. By the end of July 1967 the PoWs and the bodies of the dead had been returned to Israel.

During the War of Attrition the squadron attacked Egyptian and Syrian artillery batteries and other military facilities and terrorist targets in Jordan and Lebanon. The squadron also continued to fly photo-reconnaissance missions over the front lines. With the end of the War of Attrition in July 1970 the Vautour was approaching the end of its operational service. The 'Knights of the North' had ceased flying the Vautour by the end of 1970 and began preparations to receive the new A-4.

On 2 April 1971 the squadron received its first A-4s and soon found itself participating in numerous attack missions on targets in Syria and Lebanon. The October 1973 War found the 'Knights of the North' at full strength, with 32 aircraft and 62 pilots. The squadron flew 1136 operational sorties, mostly against Egyptian and Syrian targets. The squadron lost 11 aircraft and six pilots, including Lt Col Lev Arlozor – the commander of Ramat David – who was flying a ground-attack mission in Egypt when he

Chapter 11

Taking off from Ramat David for a training sortie, F-16C serial number 393 carries an inert Sidewinder on the wingtip, an inert GP bomb to starboard, and a Rafael Spice acquisition round to port.
(Ofer Zidon)

Also equipped with a Spice training round, F-16C serial number 359 prepares to take off from Ramat David for a training sortie.
(Ofer Zidon)

157

was shot down by AA fire and lost at sea. One of the unit's pilots ejected safely on no fewer than three occasions during the war. The squadron received 10 replacement A-4s direct from US Navy units. These were quickly painted in IDF/AF colours and joined the squadron for the last days of fighting.

In the mid-1980s the IDF/AF decided to purchase a new platform, the advanced F-16 lightweight fighter. The 'Knights of the North' was chosen to be one of the squadrons to convert to the new aircraft.

On 28 September 1980 the F-16 era officially commenced, with the arrival of two aircraft. By the end of 1980 the squadron had 20 F-16A/Bs on strength. The first operational ground-attack mission by any F-16 took place on 22 April 1981 when four aircraft from the squadron attacked a terrorist camp in Lebanon.

Operation Opera, the destruction of the nuclear reactor in Iraq, took place on 7 June 1981 with participation by eight F-16As from Ramat David: four from the 'Knights of the North' Squadron and four from the 'First Jet' Squadron. They received top cover from six F-15s.

Another first was achieved on 14 July 1981 when the squadron commander, Amir Nachumi, shot down a Syrian MiG-21. It was the first time an F-16 had shot down another jet fighter anywhere in the world.

The June 1982 First Lebanon War found the F-16s of the squadron conducting both ground-attack and interception sorties. During the first stage of the war the squadron flew 283 operational sorties, claiming 19 Syrian MiGs shot down.

The latest incarnation of the 'Knights of the North' Squadron took place on 30 July 1987 with the arrival of seven F-16C aircraft. This latest version of the F-16 featured a more powerful engine and advanced avionics, resulting in extended range and an increased payload. A few months after its arrival the new aircraft participated in a ground-attack mission over Lebanon, and over 5,000 operational sorties were conducted during 1987 and 1988. The squadron participated in Operations Accountability in 1993, Grapes of Wrath in 1996, Defensive Shield in 2003, the Second Lebanon War in 2006 (when one of the squadron's F-16s shot down a Hezbollah Ababil type UAV) and Operation Cast Lead in December 2008. During this period the squadron received new systems and new types of armament including the Python 5 AAM, Spice guided bomb and JDAM.

No. 117 'First Jet' Squadron

No. 117 'First Jet' squadron was formed on 23 June 1953 at Ramat David, to become the first IDF/AF squadron to operate jet fighters, in the form of the Meteor. The first two examples of the Meteor T.Mk 7 two-seat trainer variant arrived two weeks later and were given individual names – *Sufa* (storm) and *Sa'ar* (tempest) by Israel's first Prime Minister, David Ben Gurion.

The first two Meteor F.Mk 8 fighters arrived in August 1953. Since the Meteors represented the cutting edge of the fighter force, one or two pairs were often detached to Hatzor for QRA duties during 1954 and 1955. More aircraft arrived in early 1955, in the shape of seven ex-RAF Meteor FR.Mk 9s. Only two of these were used for their original photo-reconnaissance purposes, while the other five were used for air superiority missions.

Chapter 11

F-16C serial number 353 leaves it HAS at Ramat David, armed with a JDAM, Rafael Python 5 and AIM-9 Sidewinder AAMs.
(Ra'anan Weiss).

F-16C serial number 341 armed with Rafael Spice and Python 5 weapons.
(Ra'anan Weiss)

On 1 September 1955 a pair of Meteors from the squadron detachment at Hatzor shot down two Egyptian de Havilland Vampire fighters that were strafing a kibbutz in the Negev. These were the first jet kills for the IDF/AF. From summer 1956, with the arrival of the new Mystère IVA, the 'First Jet' Squadron lost its place as the premier interceptor unit of the IDF/AF and became the advanced training unit of the Flight School, with ground attack becoming the squadron's secondary mission. The Meteors received a coat of camouflage paint, to reflect their change in role.

During the 1956 Sinai Campaign the squadron was deployed to Tel Nof. The opening move of the war, on 29 October 1956, saw Meteor F.Mk 8 and FR.Mk 9 jets from the squadron escorting Dakota transports dropping the Paratroopers Brigade over Sinai. The next day and subsequently, the Meteors were used primarily for ground-attack missions, being particularly effective against Egyptian armoured vehicles. A total of 77 sorties were flown during the campaign, including a few photo-reconnaissance flights, for no loss.

In mid-1956, the squadron was planned to receive some 24 Canadair Sabres, but the purchase was cancelled due to political pressures. During 1957, four Belgian Meteors were purchased and flown to Israel. These aircraft were a mix of T.Mk 7 fuselages and F.Mk 8 tail units, and were known as the 'Meteor 7.5' in IDF/AF service. On 11 February 1962 the squadron was officially decommissioned, and its T.Mk 7 aircraft were transferred to No. 110 'Knights of the North' Squadron, to assist in the training of Vautour pilots, while its F.Mk 8/FR.Mk 9s were transferred to No. 107 'Orange Tail' Squadron. A few months later, on 7 July 1962, the squadron received the first Mirage IIICJ, to become the first squadron operating Mach-2 fighters in the IDF/AF.

The first day of the June 1967 War saw Mirages of the 'First Jet' Squadron leading the first wave of attacks on Egyptian air bases at Beni Suef and Faid. Raids on Syrian and Jordanian positions followed. The Mirages returned to the air superiority mission on the second day of the war. During the war the squadron flew 424 sorties, claiming a total of 12 kills while losing three Mirages, one in an aerial battle over H-3 air base in Iraq and two to ground fire, with two pilots killed and a third ejecting safely.

During the 1969–70 War of Attrition the squadron returned to its primary role of interception, taking part in a large number of aerial battles over Syria and Egypt. The final act of the war was the dogfight against MiG-21s flown by Soviet pilots. Four of the squadron's Mirages participated in a well-planned ambush that resulted in the shooting down of five Soviet pilots with no loss to the Israeli side.

With the arrival of the A-4 and F-4 the Mirages gradually became the primary interceptor of the IDF/AF. When the October 1973 War broke out, the 'First Jet' Squadron primarily flew interception and escort missions. In the 19 days of fighting the squadron flew 716 operational sorties, and claimed 55 enemy MiGs as shot down for the loss of four Mirages: one pilot was killed, two pilots ejected and were captured by the Syrians, and one ejected over the sea and was rescued.

The 'First Jet' Squadron continued flying its Mirages until 21 October 1979 when the squadron was officially disbanded. Nine months later, on 2 July 1980, the first four F-16As landed at Ramat David to reactivate the 'First Jet' Squadron as the first F-16 squadron within the IDF/AF. On 28 April 1981 a Syrian Mi-8 helicopter was shot down, to become the world's first F-16 kill.

Chapter 11

F-16C serial number 350 takes off from Ramat David. (Ofer Zidon)

F-16C tail number 337. Note the inscription at the base of the tail marking the 60th anniversary of the squadron. (Ofer Zidon)

161

On 7 June 1981 four of the squadron's F-16As flew a distance of 1,000km (625 miles) on a mission to attack the Iraqi Osirak nuclear reactor near Baghdad. The attack was completed successfully and all aircraft returned safely.

During the June 1982 First Lebanon War the squadron flew air-to-air missions, claiming 19 Syrian MiGs without loss. The squadron set a number of records in that period of time: one of the squadron's pilots claimed two MiG-23s with AAMs and one MiG-21 with cannon within 45 seconds, and on the last day of the war a four-ship formation claimed nine MiGs killed in one sortie (no. 1 – two kills, no. 2 – one kill, no. 3 – two kills and no. 4 – four kills).

During 1987 the squadron's F-16A/Bs were replaced by the more advanced F-16C/D variant. These aircraft serve with the squadron to this day, and have taken part in numerous conflicts over the last 25 years. Between 2010 and 2012 the squadron's F-16Cs underwent the Barak 2020 improvement programme aimed at extending their operational lifespan for another 10 years or more.

No. 193 'Defenders of the West' Squadron

No. 193 'Defenders of the West' Squadron is the maritime helicopter unit of the IASF. The squadron's operations budget comes from the Israeli Navy, as do the systems operators on board the helicopters.

Two HH-65 Dolphin helicopters were purchased from the US in 1985 as a stopgap solution to the Israeli Navy's requirement to expand its corvettes' detection range, in line with the purchase of the long-range Harpoon anti-ship missile. At first the helicopters were operated as a flight within No. 124 'Rolling Sword' Squadron, which otherwise operated the Bell 212. However, in August 1987 the 'Defenders of the West' Squadron was officially established. The squadron's missions comprised detection of targets beyond the radar range of the Navy's ships and the horizon, anti-submarine warfare, maritime search and rescue, and sea to shore transportation. The helicopters initially operated from a platform added to the Navy's Sa'ar 4.5 corvettes.

AS.565MA serial number 889 demonstrates the use of the winch for rescue missions during a display at its Ramat David home base.
(Ofer Zidon)

Chapter 11

The squadron's first operational assignment took place in April 1989, during a search for terrorists attempting to reach Israel's shores from the sea, together with a Seascan maritime patrol aircraft and the Navy ship *Geula*.

The squadron was closed temporarily due to budget cuts in 1990 and reactivated in April 1995 at Ramat David. In August 1996 the squadron was re-equipped with five AS.565MA Panther helicopters, a more powerful and advanced helicopter than the HH-65. The new helicopters introduced a self-protection suite that included electronic warfare systems and chaff/flare dispensers. It was also fitted with a forward-looking infrared (FLIR) turret to aid its main mission of surveillance and detection of enemy vessels.

On 16 September 1996, during a night training sortie, one of the squadron's two Dolphin helicopters crashed, claiming the lives of the squadron commander, one of the squadron's pilots and a Navy systems operator.

Through the years, the squadron's helicopters have taken part in many operational activities that have helped defend Israel's coastline – the country's longest border. They have also operated far from Israel's shores against ships en route from Lebanon and Egypt, attempting to smuggle weapons to terrorists in Gaza Strip, as typified by the capture of the ship *Santorini* in 2001.

During the Second Lebanon War in 2006 Hezbollah launched a Chinese-made C-802 coastal defence missile against the Israeli Navy corvette *Hanit*. The missile hit the ship's stern, killed four soldiers and damaged the Panther helicopter on board. The helicopter was returned to service after repairs in 2008.

AS.565MA serial number 891 is restrained on the flight deck of an Israeli Navy Sa'ar 5 corvette. (IDF Spokesperson)

Modern Israeli Air Power

Panther serial number 895 during refuelling from a Navy vessel.
(IDF Spokesperson)

Panther serial number 885 lands on board a Navy Sa'ar 5 corvette.
(IDF Spokesperson)

RAMON AIR BASE

Brief history

Ramon air base was built in the Negev, near the Ramon crater. After Israel's withdrawal from its air bases in Sinai under the terms of the peace agreement with Egypt, Ramon was built to replace Eitam in Sinai. The new base was activated on 19 May 1982 with the arrival of the Skyhawk-equipped No. 140 'Golden Eagle' Squadron from Etzion in Sinai, and the F-16A/B-equipped No. 253 'Negev' Squadron from Eitam in Sinai. The 'Negev' Squadron flew Mirages at Eitam and was converted to the F-16 prior to its transfer to Ramon.

Two years later, in 1984, Ramon saw the arrival of its first helicopter unit, No. 190 'Magic Touch' Squadron equipped with the Model 500 Defender. The 'Magic Touch' Squadron remained at the base until 1990 when it was transferred to Ramat David for a period of five years. In 1995 the squadron returned to Ramon as the second AH-64 unit at the base.

The first AH-64 squadron was No. 113 'Hornet' Squadron that was activated in 1990. This marked the first time that a fighter squadron had been re-formed as a rotary-wing squadron. During the 2000s Ramon became the most technologically advanced Israeli Air Force base, with the arrival of the AH-64D attack helicopter and the most sophisticated Fighting Falcon variant, the F-16I. Currently the base hosts five squadrons: two AH-64 squadrons (No. 113 'Hornet' and No. 190 'Magic Touch') and three F-16I squadrons (No. 119 'Bat', No. 201 'The One' and No. 253 'Negev'). All of these squadrons took part in the conflicts in Gaza and Lebanon fought during the last decade.

No. 113 'Hornet' Squadron

No. 113 'Hornet' Squadron was established on 4 October 1955, equipped with the new Ouragan. The second type of jet fighter in the IDF/AF, after the Meteor, the Ouragan marked the beginning of the French era in the force. The aircraft arrived from two French Air Force operational squadrons, drawn from two fighter wings: Escadre de Chasse 12 and Escadre de Chasse 4. This fact explains the origins of the 'Hornet' Squadron insignia and its resemblance to the insignia of the 2e Escadrille of Escadron de Chasse 1/12 'Cambrésis'. The squadron received 24 Ouragans that represented the cutting edge of the IDF/AF until the arrival of the Mystère IVA. On 12 April 1956 one of the squadron's Ouragans shot down an Egyptian Vampire (plus one unconfirmed) that were patrolling the Negev area.

Modern Israeli Air Power

AH-64D serial number 113 is unloaded from a Russian Antonov An-124 transport aircraft. The An-124 brought the three most recently upgraded AH-64Ds to Israel. The original AH-64As had been sent to the US a year earlier. These rebuilt airframes joined No. 113 'Hornet' Squadron.
(Ofer Zidon)

Former AH-64A serial number 929, AH-64D serial number 113 received its new identity to mark its service with No. 113 'Hornet' Squadron. Many IASF squadrons have one aircraft that carries the unit's number.
(Ofer Zidon)

Chapter 12

AH-64D serial number 739 over Mount Hermon in northern Israel. The Saraf is armed with six Hellfire missiles. (Ofer Zidon)

During the Suez campaign, from 29 October 1956, Ouragans attacked Egyptian ground positions and armoured vehicles in the Sinai. They were also involved in the capture of the Egyptian destroyer *Ibrahim el-Awal* on 31 October. A total of four Vampires and one MiG-15 were claimed shot down by Ouragans during the campaign, for the loss of two Ouragans. With the arrival of the Mystère IVA the Ouragan lost its front-line interceptor position and the 'Hornet' Squadron became an OTU in 1957.

The 'Hornet' squadron was still operating the Ouragan during the June 1967 War. The Ouragans flew 450 operational sorties against Egyptian airfields in the Sinai on 5 June, followed by raids on the Syrian and Jordanian fronts for the remainder of the war. The 'Hornet' Squadron is known to have shot down one MiG-21 while eight Ouragans were written off in return.

The War of Attrition saw the amalgamation of the Ouragans from No. 107 'Orange Tail' Squadron within the 'Hornet' Squadron, and the type resumed its OTU role until 1972 when it was replaced by A-4 in this role and retired from service. During the fighting, two Ouragans were lost during an attack mission over Jordan on 21 March 1968,

167

Modern Israeli Air Power

while missions against Egypt were also flown. In November 1972, with the retirement of the Ouragans, the squadron converted to the Nesher. During the October 1973 War the squadron flew its Neshers in some 660 operational sorties in the interception and air superiority role, claiming over 50 MiGs kills for no loss.

In 1976 the squadron converted again, this time to the Nesher's successor, the Kfir C1, becoming the second unit to operate the new type. The 'Hornet' Squadron performed the first ever ground-attack mission with the Kfir on 9 November 1977, striking PLO positions at Tel Aziyah in Lebanon. On 21 November 1977, four of the squadron's aircraft escorted the Boeing 707 of Egyptian President Anwar Sadat back home after his historic visit to Israel.

In 1979 the more capable Kfir C2 replaced the earlier version. The type was used effectively during the First Lebanon War in June 1982, including attacks on Syrian SAM radars and on PLO positions around Beirut. The squadron was disbanded in 1987, due to budget cuts and the phasing out of the Kfir from service.

The 'Hornet' Squadron was reformed in September 1990 at Ramon, to become the IDF/AF's first AH-64 attack helicopter operator. In April 2005 the 'Hornet' Squadron took delivery of the advanced AH-64D-I Apache Longbow, the type it flies to this day,

AH-64D serial number 758 undergoes maintenance. The ground crew is arming the helicopter with Hellfire missiles. (Ofer Zidon)

Chapter 12

and transferred its older AH-64As to No. 190 'Magic Touch' Squadron. During its AH-64 period of operations the squadron has participated in thousands of operational sorties over the Gaza Strip and Lebanon, most of them against high-value targets that require the accuracy of the helicopter's AGM-114 Hellfire missile. A significant number of these attacks were 'targeted killings' of terror leaders and operatives. One of the 'Hornet' Squadron's AH-64Ds was lost during the 2006 Second Lebanon War due to a technical malfunction.

No. 119 'Bat' Squadron

No. 119 'Bat' Squadron was formed on 9 August 1956. Initially, it operated three Meteor NF.Mk 13s from Ramat David and became the IDF/AF's first all-weather and night fighter squadron. The first operational mission for the new squadron was of great significance. On the night of 28 October 1956, one day before the start of the Sinai Campaign, a Meteor NF.Mk 13 shot down an Egyptian Il-14 en route from Damascus to Cairo, with many high-ranking Egyptian officers killed. The squadron's successful

Four F-16Is from No. 119 'Bat' Squadron stand guard at the end of Ramon's runway. (Ofer Zidon)

169

Modern Israeli Air Power

F-16I serial number 451 prepares to take off fully loaded with a GBU-10 laser-guided bomb, Python 5 and AIM-120. (Ofer Zidon)

operations during the Sinai Campaign led to the decision to transfer the squadron to Tel Nof, closer to the border with Egypt – at that time the main threat to Israel. The squadron moved to its new home on 15 December 1957.

Because of its experience in night-time operations, the 'Bat' Squadron was chosen to receive the new Vautour IIN, a version adapted for adverse-weather and night operations. The Vautours served with the squadron from July 1958 until July 1963, when the squadron converted to the Mirage III. The 'Bat' Squadron retained its photo-reconnaissance capability with two specially equipped Mirages.

During the October 1967 War the 'Bat' Squadron flew over 300 operational sorties, mostly air-to-ground missions, and claimed a total of 23 MiGs shot down.

In 1970 the 'Bat' Squadron became the third F-4 squadron in the IDF/AF. During the October 1973 War the squadron flew almost 1,000 operational sorties against ground targets in Egypt and Syria. Among them were long-range attacks over Egypt and 162 photo-reconnaissance missions. The squadron's most famous mission of the war was the attack on the Syrian Army HQ in Damascus. Because of bad weather conditions, the 'Bat' Squadron was the only unit to complete the mission, from the three F-4 squad-

Chapter 12

F-16I serial number 463 at Ramon. The Sufa is on alert, armed with a Rafael Spice guided munition, Python 5 and AIM-120 AAMs.
(Ofer Zidon)

rons sent for the mission. Three of the squadron's pilots received citations for continuing with the mission despite difficult conditions and great risk. During the war the squadron claimed 14 enemy MiGs destroyed and lost five of its Phantoms, with four aircrew killed and four becoming PoWs. The squadron's crews claimed an additional Iraqi MiG-21 that supposedly crashed while chasing a pair of RF-4Es over western Iraq after the latter took reconnaissance photos of H-3 air base. During the June 1982 First Lebanon War, 18 of the squadron's Phantoms participated in the attack of the Syrian SAMs in Lebanon. On 24 July 1982 one of the squadron's aircraft was shot down with one pilot killed and the second becoming a PoW in Syria.

On 28 December 2004 the 'Bat' Squadron was reactivated to become the second F-16I squadron in the Israeli Air Force and at Ramon. During its first year, in 2005, the squadron flew 39 operational sorties, mainly ground-attack missions over the Gaza Strip.

In the last decade the 'Bat' Squadron has participated in all Israeli conflicts including the Second Lebanon War, when one of its F-16Is was destroyed during take-off because of a main wheel failure. The squadron flew 200 operational sorties during Operation Cast Lead in late 2008.

No. 190 'Magic Touch' Squadron

No. 190 'Magic Touch' Squadron was activated at Palmachim as a dedicated operator of the Model 500 Defender. Prior to the June 1982 First Lebanon War the squadron was transferred to Ramat David in order to reduce the distance to its main theatre of operations on the Lebanese border. The young squadron was heavily involved in the war, flying numerous attack and close air support sorties against Syrian Army units in Lebanon. On 8 June 1982 the Defenders attacked and destroyed a Syrian radar station in preparation of the IDF/AF attack against the Syrian SAM network in Lebanon. The next day saw the Defenders taking part in a battle against Syrian armoured units, with better results than the more sophisticated AH-1s. During the war one of the squadron's helicopters was shot down and its crew was rescued by a Bell 206. In August 1984 a pair of Defenders attacked a terrorist base 170km (105 miles) into Lebanon, after taking off from the rear deck of an Israel Navy Sa'ar corvette.

In 1984 the squadron was transferred to Ramon, but with increasing tension in the north it returned to Ramat David in 1990. The 'Magic Touch' Squadron participated in Operation Accountability in 1993, attacking 20 terrorist targets and tanks in Lebanon.

A year later, in 1994, the squadron was decommissioned and the Defender was retired in favour of additional AH-1s and the purchase of a second batch of AH-64s. The squadron was reactivated in March 1995 at Ramon as the second Israeli AH-64 squadron. With the arrival of the AH-64D in 2005, No. 113 'Hornet' Squadron transferred its AH-64As to the 'Magic Touch' Squadron, which doubled its strength. All the remaining AH-64As now remain in service with the 'Magic Touch' Squadron, as the aircraft upgraded to AH-64D standard have been passed on to Hornet' Squadron.

During the last decade the squadron has taken part in operational activities in Gaza and Lebanon together with the 'Hornet' Squadron.

AH-64A serial number 924 employs its 30mm cannon during a display for the Air Force Day and graduation ceremony.
(Ofer Zidon)

Chapter 12

AH-64A serial number 832 releases flares during a graduation ceremony at Hatzerim.
(Ofer Zidon)

Armed with Hellfire missiles, a pair of AH-64As fly over the Golan Heights.
(Ofer Zidon)

No. 201 'The One' Squadron

No. 201 'The One' Squadron was established on 17 August 1969 in Hatzor as the IDF/AF's first F-4 squadron. The squadron was activated in the middle of the War of Attrition. Since the F-4 was far more advanced and better suited to air-to-ground operations than any other IDF/AF aircraft, the squadron began operations only two weeks after its activation. The introduction of the Phantom allowed the IDF/AF to increase its range of activities and to include long-range strategic strikes against targets deep inside enemy territory. On 11 November 1969 the Israeli Phantom claimed its first kill, against an Egyptian MiG-21. The main task of both Phantom squadrons was strikes against the extensive SAM network that had been deployed by the Soviets between Cairo and the Suez Canal. On 18 July 1970 'The One' Squadron and its sister unit, No. 69 'Hammers' Squadron, attacked the SAM network at great cost to themselves. The commander of 'The One' Squadron was hit by an SA-3 missile and was killed, while the commander of the 'Hammers' was also hit but was able to perform an emergency landing in his damaged Phantom.

F-16I serial number 854 armed with a Rafael Spice guided munition and Python 5 and AIM-120 AMRAAM missiles. (Ofer Zidon)

Chapter 12

A trio of F-16Is lines up on the Ramon runway, waiting for take-off clearance.
(Ofer Zidon)

F-16I serial number 890 armed with the Israel Military Industries (IMI) Delilah missile, Python 5 and AIM-120.
(Ofer Zidon)

175

Modern Israeli Air Power

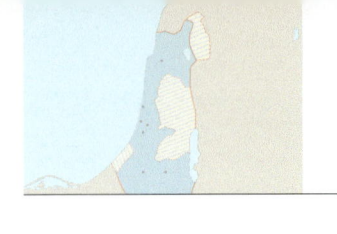

F-16I serial number 878 displays a typical Quick Reaction Alert interception loadout, consisting of wingtip AIM-120s and Python 4/5 AAMs inboard, plus a centreline drop tank and Litening pod. (Ofer Zidon)

During the October 1973 War the squadron flew 758 operational sorties, most of them hazardous long-range attacks and missions against SAM batteries. The squadron claimed 32 enemy aircraft shot down while losing 14 Phantoms, most of them to Syrian SAM sites. Seven aircrew were killed and 14 became PoWs. During Operation Litani in 1978 the squadron flew close air support and interception patrols. During the First Lebanon War in 1982 the squadron played a major role in the attack on the Syrian SAM network in Lebanon, without sustaining any casualties. 'The One' Squadron also flew interception patrols and close air support missions.

On 19 January 1988 the squadron was transferred to Tel Nof and a year later, on 9 April 1989, it was the first squadron to convert to the Kurnass 2000, an Israeli upgrade of the F-4. 'The One' Squadron entered the world of photo-reconnaissance missions using new and improved photography equipment.

'The One' Squadron was reactivated on 9 July 2008 as the last of the IASF's four F-16I squadrons. The reactivation took place at Ramon, where 'The One' Squadron joined the 'Bat' and 'Negev' Squadrons already flying the F-16I from the base.

No. 253 'Negev' Squadron

No. 253 'Negev' Squadron was formed on 27 July 1976 at Hatzor. It was initially equipped with the Nesher. In October 1976 the squadron was transferred to Eitam in Sinai. Two years later, the squadron flew 26 patrol and interception missions during Operation Litani. During 1979 the Neshers were sold to Argentina and on 14 June 1979 the squadron was re-equipped with Mirage IIIs from No. 117 'First Jet' Squadron, which in turn became an F-16A operator. Following the signing of the Israeli-Egyptian peace treaty, Israel withdrew from Sinai and Eitam was closed. On 25 April 1981 the squadron's Mirages were transferred to No. 254 'Midland' Squadron, a reserve unit established at Hatzor. Later that year, in October 1981, the squadron was reactivated at Ramat David

Chapter 12

to become the third F-16A/B squadron of the IDF/AF, and in February 1982 it was transferred to its current home base at Ramon, which had been built as replacement for Eitam in Sinai.

During the First Lebanon War the squadron flew 192 patrol, intercept and escort missions. During Operation Accountability in 1993 it flew 20 sorties, comprising eight air strikes and two patrol and interception sorties. In 1995 the squadron received a number of surplus USAF two-seat F-16Bs and began the training and integration of navigators as part of its regular operations. On March 2003 the squadron transferred some of its F-16A/Bs to Nevatim-based No. 116 'Defenders of the South' Squadron in preparation to become the IDF/AF's first F-16I outfit. The transition took place on 19 February 2004, with the arrival of the first pair of the latest fighter in the Israeli inventory. During the last decade No. 253 'Negev' Squadron has been involved in numerous operations, including the 2006 Second Lebanon War and Operation Cast Lead in 2008.

F-16I serial number 407 taxis from its HAS, armed with a GBU-10 bomb, Python 5 and AIM-120 AMRAAM. (Ofer Zidon)

177

Modern Israeli Air Power

F-16I serial number 444 performs a tight turn to reveal the upper wing and fuselage camouflage scheme. (Ofer Zidon)

F-16I serial number 456 in the standard air-to-air configuration of AIM-120 and Python 5 missiles. (Ofer Zidon)

Chapter 12

F-16I serial number 404 takes off from Ovda runway, during one of the squadron's deployments to the Advanced Training Center.
(Ofer Zidon)

F-16I serial number 415 is seen armed with GBU-15 bombs.
(Courtesy 'Negev' Squadron)

179

SDE DOV AIR BASE

Brief history

Sde Dov is located in the northern outskirts of Tel Aviv and hosts IASF activities together with civil aviation. The IASF base and airport is named after Dov Hoz, one of the pioneers of Jewish aviation. The airfield was built in the late 1930s, in the middle of the Arab revolt against the British mandate in Palestine. Israel Rokach, the mayor of Tel Aviv, requested the British authorities for permission to create an airport in Palestine in order to solve the transportation problem of Jews during the Arab revolt. The airfield would also serve the needs of the Riding power plant that was built at that time. Work began on a plot of land north of Tel Aviv in 1938. The new airport served regular flights to Haifa, with the option of flights to Beirut.

During the 1940s Sde Dov was the base for activities by Aviron, a Jewish aviation company that was a front for the Hagana (the Jewish military establishment that preceded the IDF) and the predecessor of the Israeli Air Force. During the 1948 War of Independence the base hosted Tayeset Alef (Squadron A), which consisted of two dozens light aircraft including the Autocrat and the Polish-built RWD-13. The light aircraft were used for liaison, transport, intelligence gathering and even bombing missions against Arab forces that attacked the Israeli towns. On 15 May 1948, with the outbreak of the war, Sde Dov was damaged in an attack by four Egyptian Spitfires. At the time Tayeset Alef became No. 100 'Flying Camel' Squadron, which is still based at Sde Dov today. Over the years the 'Flying Camel' Squadron has flown various types of light transport including the Cessna 206, Dornier Do 27 and Do 28. Today it flies the King Air 200.

In 1967 No. 125 'Light Helicopter' Squadron was activated at Sde Dov. Over the years the squadron flew the Bell 47, Alouette II, Bell 206 and 206L in transport, liaison and ground support missions. The squadron was decommissioned in 2003.

On 1 May 1974 No. 135 'Light Transport' Squadron was established at Sde Dov. Since the 'Flying Camel' Squadron grew in strength after the October 1973 War, it was decided to form a second squadron. The new squadron assumed some of the missions of the 'Flying Camel' Squadron, including operational training for new transport pilots and navigators. Today the squadron flies three types of Beechcraft aircraft: the King Air 200, RC-12 (the military version of the King Air) and the Model 36 Bonanza. In 2004 the unit's name was changed accordingly to the 'Kings of the Air' Squadron.

Both squadrons fly the highest number of sorties in the IASF. Most of these are operational intelligence-gathering missions, using specialised observation systems on board the aircraft. Despite this, the days of Sde Dov as an active military air base are numbered. As part of the wave of cutbacks announced in 2013, an agreement was signed on 14 August 2013 between the ministries of finance, transportation and housing regarding the closure of Sde Dov, and the construction of 16,000 apartments on the site. The base will therefore close in 2018 – or perhaps even as early as 2016. Civil transportation will be transferred to Ben Gurion Airport while the military activities will be transferred to Haifa.

No. 100 'Flying Camel' Squadron

No. 100 'Flying Camel' Squadron was officially formed in January 1949 at Tel Nof and gathered all the light aircraft of the young IDF/AF. Its roots lay in the aviation activities of the Jewish administration in Palestine, during the British mandate. The Aviron company was formed during the 1940s as a front for Jewish military aviation activities. In 10 November 1947 Tayeset Alef (Squadron A) was formed in Sde Dov. The war between the Jews and Arabs in Palestine had already begun and many Jewish towns were under siege or subject to attacks. The only way to transport supplies and manpower and extract wounded from these towns was by air. Tayeset Alef inherited 11 light aircraft from Aviron and received 21 surplus British Autocrats. This fleet carried the burden during the first months of the War of Independence. The squadron's missions included transport, evacuation, liaison, reconnaissance and photography and even bombing – the crew threw small bombs from the Auster's windows. The squadron was divided into three flights, one operating in the south of Israel, one in the north, and one from Sde Dov in central Israel.

During the 1956 Sinai Campaign the squadron's Piper Cubs flew 169 operational sorties in support of ground forces, observation and light transport. One of the squadron's Pipers rescued Benny Peled, who later became IDF/AF commander, after his Mystère was hit and he was forced to eject. Five of the pilots received citations for their activities during the war. After the war the squadron received the more powerful Piper Super Cub.

King Air serial number 856 takes off from Sde Dov. Note the civilian colour scheme.
(Ofer Zidon)

Chapter 13

King Air serial number 633 takes off from Sde Dov. Note the surveillance pod under the fuselage.
(Ofer Zidon)

King Air serial number 735 on the Sde Dov runway, which runs parallel to the Tel Aviv beach.
(Ofer Zidon)

Modern Israeli Air Power

In April 1959 the squadron returned to Sde Dov, which remains its home base today. During the June 1967 War the squadron flew a total of 314 operational sorties: 167 scouting and observation sorties, 136 transport sorties and 11 communications relay sorties.

After the war No. 100 'Flying Camel' Squadron was strengthened with a number of new Cessna 206 Skywagons in 1968 and Do 27s and Do 28s in 1971, and was able to hand over its Pipers to the Flight School.

The squadron flew 242 operational sorties during the October 1973 War. The squadron's fleet saw the addition of 14 Britten-Norman Islander light transports recruited after the war broke out. After the war the squadron was divided with the establishment of a second squadron in Sde Dov. This was No. 135 'Light Transport' Squadron, which received the Islander and Cessna 206 aircraft from the 'Flying Camel' Squadron, together with their light transport missions. No. 100 'Flying Camel' Squadron retained its Do 27 and Do 28 aircraft and their observation and reconnaissance missions, until they were replaced by the King Airs. The new type assumed the observation and scouting missions and was equipped with specialist surveillance and intelligence-gathering systems.

No. 135 'Kings of the Air' Squadron

No. 135 'Kings of the Air' Squadron was activated on 1 May 1974 as a result of lessons learned in the October 1973 War – primarily the requirement for a viable all-weather light transport force. The growth of No. 100 'Flying Camel' Squadron was another catalyst for the creation of the new unit, which assumed the 'Flying Camel' Squadron's transport and OTU missions.

The squadron was initially equipped with the Beechcraft Queen Air 80, Islander and the Cessna U-206, the last two types having been transferred from the 'Flying Camel' Squadron.

In November 1984 the squadron received the RC-12, designed for electronic warfare missions. The new aircraft broadened the squadron's scope of roles into the arena of

Bonanza serial number 304 belongs to No. 135 'Kings of the Air' Squadron, but is usually operated by Ramon commanders as indicated by the Ramon air base insignia on its tail.
(Ofer Zidon)

184

Chapter 13

RC-12 serial number 980 on approach to Sde Dov.
(Ofer Zidon)

King Air serial number 711 from No. 135 'Kings of the Air' Squadron.
(Ofer Zidon)

Seen at Ramon, Bonanza serial number 320. This aircraft displays a unique combination of civil colour scheme, US registration, 'Kings of the Air' Squadron insignia and IASF serial number. (Ofer Zidon)

operational and combat support. The squadron flew many operational sorties during the conflicts in the 1990s and 2000s, together with its traditional light transport and liaison missions.

In 1995 the Cessna 206 was retired in favour of the TB-20 Trinidad, although this turned out to be a problematic aircraft with high maintenance costs, and was in turn replaced in 2004 by a third Beechcraft type, the Model 36 Bonanza.

A major advance occurred in 2002 with the introduction of the King Air 200. The new aircraft became the primary aircraft of the light transport force and led to the retirement of many other types, including Queen Air, Do 28, Dakota and Arava. The King Air allowed the unification of the light transport force, making it more efficient and cost effective.

All of the transport stream graduates from the Flight School continue their operational training in the 'Kings of the Air' Squadron. At the end of the OTU they begin the qualification process for the light transport aircraft: King Air 200, RC-12 or Bonanza. Some of the aircrew will then continue to the heavy transport wing, where they will become aircrew on the Boeing 707, C-130 or Gulfstream aircraft.

Due to its surveillance work and other operational missions, the squadron flies the largest number of hours among the manned aircraft squadrons of the IASF.

No. 249 'Elad' Squadron

With the tragic results of the December 2010 Mount Carmel fire in mind, the Israeli government decided to create a new firefighting squadron within the IASF: No. 249 'Elad' Squadron (named after 16-year-old Elad Riven, a volunteer firefighter who lost his life fighting the Carmel blaze).

The squadron's AT-802F firefighting are operated under contract by Elbit Systems, which supplies flight hours to the IASF. The aircraft are actually flown by Chim-Nir, a specialist crop-dusting operator.

The first three aircraft arrived in Israel on 31 March 2011 and the others joined them a week later. The new firefighting squadron was declared operational on 12 May 2011.

The Air Tractor fleet is divided between Sde Dov and Megiddo airstrip near Mount Carmel. While the latter is often subject to forest fires, Sde Dov is a few minutes' flying time from the mountains of Jerusalem, another area prone to forest fires.

Chapter 13

Three AT-802Fs with conventional undercarriage and two float-equipped AT-802 Fireboss aircraft. The floats were subsequently removed after they were found unsuitable for local operations.
(Ofer Zidon)

Since their arrival the Air Tractors have tackled numerous fires during the hot and dry Israeli summertime.
(Ofer Zidon)

Chapter 14

TEL NOF AIR BASE

Brief history

Tel Nof was the main air base of the RAF during the period of the British mandate in Palestine. The base was established in 1941 near the Arab village of Aqir. The Jewish institutions in Palestine began to discuss the possibility of taking control over parts of Britain's aviation infrastructure in Palestine and on March 1948 the land on which the base was located was purchased. After a battle with Arab forces, the Hagana (the military force of the Jewish establishment in Palestine) took over the base.

During the first months of the Israel's War of Independence in 1948, transport aircraft made their way from Czechoslovakia and landed at the base to supply to the young Israel Defence Force, including dismantled S-199 fighters. Four of S-199s were assembled and flew their first operational mission on 29 May 1948. On 71 August 1948 the base was formally opened in a ceremony and its name was changed to Tel Nof after the name of a city planned to be built near the base.

During its first two decades of operations Tel Nof, the IDF/AF's largest air base, saw considerable activity. The heavy transport unit No. 103 'Flying Elephant' Squadron operated from Tel Nof until 1974 when it was transferred to Lod. No. 109 'Valley' Squadron flew its Mosquito fighter-bombers from the base during 1951 before it was transferred to Ramat David in 1952. No. 115 'Flying Dragon' Squadron operated from Tel Nof with its photo-reconnaissance Mosquitoes and Meteors from June 1956, and with A-4s from March 1969 until its transfer to Nevatim in 1982. Tel Nof saw No. 119 'Bat' Squadron operating first its Meteors and Vautours from 1958 and later the Mirage III (from 1964) and the F-4 (1970) until it was transferred to Ramon in 2004 in order to receive the new F-16I. No. 116 'Flying Wing' Squadron (which later changed its name to 'Defenders of the South') flew its Mystères from Tel Nof from 1961 and A-4s from 1971, before transferring to Nevatim in 1982. The Flight School operated from Tel Nof between 1955 and 1966 before moving to Hatzerim. Rotary-wing squadrons also operated from Tel Nof for many years, before the light assault squadrons left the base – first to depart was No. 123 'Southern Bell' Squadron to Hatzerim in 1968, followed by No. 124 'Rolling Sword' Squadron to Palmachim in 1978. No. 160 'First Attack Helicopter' Squadron also began life at Tel Nof in 1975 before it was transferred to Palmachim in 1979. No. 201 'The One' squadron flew its F-4s from Tel Nof beginning in 1969.

The October 1973 War found the base operating a mix of seven squadrons from all IDF/AF branches: No. 119 'Bat' Squadron flying F-4s, No. 115 'Flying Dragon' and No. 116 'Flying Wing' squadrons operating A-4s, No. 114 'Night Leaders' Squadron with Super Frelons, No. 118 'Night Predators' Squadron flying the CH-53, No. 124 'Rolling

Sword' Squadron flying the Bell 205, and No. 103 'Flying Elephant' Squadron with its Noratlas transport.

The heavy assault helicopter fleet of the Israeli Air Force has been located at Tel Nof since its birth in 1967. No. 114 'Night Leaders' Squadron was the first to receive the Super Frelon heavy assault helicopter in April 1966. After the October 1973 War the unit converted to the CH-53. The second squadron – No. 118 'Night Predators' – was formed in 1970 with new CH-53. Today, both squadrons fly the CH-53 Yasur 2000 and Yasur 2025 from Tel Nof.

Israel's first three F-15s arrived at Tel Nof in December 1976 and joined the newly formed No. 133 'Twin Tail' Squadron. Their first aerial victory occurred on 27 June 1979, when four F-15s fought against a dozen Syrian MiGs and shot down four. The second F-15 squadron – No. 106 'Spearhead' – was scheduled to be activated on 6 June 1982, but the start of the First Lebanon War delayed its inauguration until 13 June. In October 1985 eight F-15s from both squadrons executed the Israeli Air Force's longest raid to date, attacking the PLO HQ in Tunis in Operation Wooden Leg, at a distance of 2,060km (1,280 miles) from Israel.

Among the lessons learned in the October 1973 War was the need for a professional flight-test unit. The Flight Test Center was established in April 1974 at Tel Nof and is responsible for testing airframes and weapons, as well as new configurations and systems.

Another lesson from the war was the need for a specialised search and rescue unit. As a result, the Airborne Rescue and Evacuation Unit was established in April 1974. The formation is known as Unit 669 and utilises both the CH-53 and UH-60 fleets for its operations.

Tel Nof also hosts the IASF's Air Maintenance Unit – AMU 22. It was activated in 1949 and is responsible for D-level maintenance. Through the years the AMU has become specialised in unique projects for the restoration of aircraft severely damaged in accidents.

The newest member of the Tel Nof squadrons is No. 210 'Eitan' Squadron, activated in 2010 in order to operate the largest and most advanced UAV in IASF service, the turboprop-powered Heron TP.

No. 106 'Spearhead' Squadron

The first incarnation of No. 106 'Spearhead' Squadron was brief. It was activated in December 1948, flying transport missions to the Israeli Negev towns that were under siege, using C-46s. The squadron took part in bombing operations during the battle over the Negev in the beginning of 1949 and was disbanded in May 1949 with the end of the War of Independence.

With the accumulation of F-15s from the second procurement contract the IDF/AF planned to activate a second F-15 squadron. The commissioning date was set for 6 June 1982, but with First Lebanon War began the same week and the opening ceremony of the new incarnation of the 'Spearhead' Squadron was postponed to 13 June 1982. The squadron was also responsible for the last two manned fighter kills to have been claimed by Israel: on 19 November 1985 its F-15s shot down a pair of Syrian MiG-23s.

Chapter 14

F-15C serial number 560 taxis after landing at Tel Nof. The aircraft's individual name is *Tzedek* (Jupiter).
(Ofer Zidon)

Since its activation the squadron has claimed 22 aerial victories over Syrian MiGs. A unique victory was gained when, in August 1982, the squadron took part in the shooting down of a Syrian MiG-25R. The aircraft was flying a photo-reconnaissance mission over southern Lebanon at an altitude of 70,000ft (21,336m) and a speed of Mach 2.5. No IDF/AF asset was able to match to that speed and altitude. On 31 August 1982 two HAWK SAMs were fired at the MiG-25 from a battery positioned high in the Lebanese mountains. The MiG was hit and began to lose altitude, falling into the hands of a pair of the squadron's F-15s, which finished the job and shot the MiG-25 down.

On 1 May 1983 one of the squadron's F-15s was involved in a mid-air collision with an A-4. The F-15 lost its entire starboard wing but was able to land safely at Tel Nof.

F-15D serial number 706. The aircraft's individual name is *Kochav Hatzafon* (north star). The fighter is armed with the Rafael Popeye guided munition under its port wing. Note the SATCOM antenna just aft of the cockpit.
(Ofer Zidon)

Modern Israeli Air Power

F-15D serial number 706 during a mock dogfight over Ovda. (Ofer Zidon)

On 1 October 1985 the squadron was involved in Operation Wooden Leg, when its F-15s bombed the PLO HQ in Tunis together with F-15s from No. 133 'Twin Tail' Squadron. After 20 years of service and with no replacement in sight, Israel decided to implement an improvement project for its ageing F-15s. Known as Improved Baz, the programme lasted between 1995 and 2005 and included many avionics improvements including GPS-based navigation and communication systems, new cockpit displays, and other changes. Currently the F-15s are undergoing a new improvement programme, Baz Forever, which includes structural enhancements and the replacement of wiring in the fighter.

Over the years, and in light of the change in the operating environment and developing threats to Israel, the F-15 has gained some air-to-ground capabilities, using Popeye TV-guided missiles and the GPS-guided JDAM series of munitions.

F-15C serial number 575 lands at Tel Nof. The aircraft carries the individual name *Ma'adim* (Mars).
(Ofer Zidon)

Chapter 14

No. 114 'Night Leaders' Squadron

No. 114 'Night Leaders' Squadron was activated on 6 January 1966 at Tel Nof, to become the IDF/AF's second helicopter squadron and its first heavy-lift helicopter unit. The squadron initially flew the Super Frelon helicopter. During the June 1967 War and the 1969–70 War of Attrition the squadron flew numerous operational sorties including the infiltration of special forces behind enemy lines, search and rescue missions, transport and liaison missions and evacuation of wounded soldiers from the front lines. On 17 October 1969 the squadron received the first pair of CH-53s, which were operated by a separate flight within the squadron. Two months later a mixed force of Super Frelons and CH-53s raided an advanced Egyptian radar station and brought it to Israel carried under the belly of the CH-53s. The squadron's CH-53s were transferred to the newly formed No. 118 'Night Predators' Squadron on 6 August 1970, with the end of the War of Attrition.

During the October 1973 War the squadron flew 331 operational sorties with its Super Frelons, including the attempt to re-conquer Mount Hermon from the Syrian Army, during which one of the helicopters crashed and six crewmembers were killed.

After the 1973 war the squadron was reinforced with a number of surplus US Marine Corps CH-53As, which were flown together with the older Super Frelons until 31 March 1991, when the latter type was retired from service.

The squadron works in close corporation with IDF infantry brigades, special forces and Unit 669, the IASF's dedicated search and rescue unit. The squadron has participated in countless operations during the last three decades including infiltration operations of special forces behind enemy lines. The squadron's helicopters, like those of the 'Night Predators' Squadron, underwent two major improvement programmes – Yasur 2000 in the 1990s that involved new avionics and navigation systems, electronic warfare and self-protection systems, new cockpit displays and new titanium rotor blades, and Yasur 2025 between 2010 and 2013 that saw structural enhancements, tail

CH-53 Yasur 2000 serial number 046 undergoes maintenance on the squadron apron in Tel Nof. (Ofer Zidon)

Modern Israeli Air Power

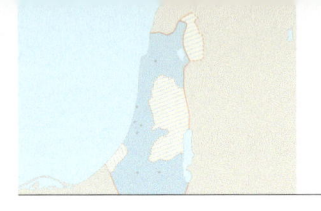

CH-53 Yasur 2000 serial number 979 flies fast and low over Hatzerim (above) before deploying special forces (right). The helicopters in the 9xx range are surplus US Marine Corps aircraft that arrived in two batches, in 1974 and 1991. Note the reinforcement rods for the external fuel tanks.
(Ofer Zidon)

Chapter 14

CH-53 Yasur 2000 serial number 065. The helicopters in the 0xx range are the original aircraft that were delivered new to Israel in the early 1970s. (Ofer Zidon)

unit replacement and more, with the target of extending the helicopter's life span until at least 2025.

During the 2006 Second Lebanon War the squadron flew hundreds of missions, including assault transport of an infantry brigade into southern Lebanon. In one of the missions a CH-53 from the squadron was hit by an anti-aircraft missile just minutes after the disembarkation of an infantry force and crashed, killing five crewmembers.

No. 118 'Night Predators' Squadron

The heavy assault helicopter fleet of the Israeli Air Force grew in number with the purchase of new CH-53s and a decision was made to divide the fleet between two squadrons. No. 114 'Night Leaders' Squadron would fly the Super Frelon while a new squadron was formed on 6 August 1970 to operate the CH-53. No. 118 'Night Predators' Squadron was inaugurated at Tel Nof and received the CH-53s flown by 'Night Leaders' Squadron, as well as the newly arrived helicopters from Sikorsky.

During the October 1973 War the squadron flew hundreds of operational sorties, encompassing assault transport, infiltration of special forces behind enemy lines, search and rescue operations, evacuation of wounded soldiers, and electronic warfare missions. The highlight of the squadron's operations was the battle for Mount Hermon, when, on 21 October 1973, the squadron's CH-53s flew 626 paratroopers on to the Syrian side of Mount Hermon at an altitude of over 2,500m (8,202ft). Their mission was to regain Mount Hermon, which had been conquered by Syrian forces in the opening stages of the war.

In the June 1982 First Lebanon War the squadron flew over 200 operational sorties including assault transport, evacuation missions and joint operations with special

Modern Israeli Air Power

A pair of CH-53 Yasur 2000s, serial numbers 921 and 912 take off from Tel Nof. (Ofer Zidon)

forces. The 'Night Predators' CH-53s also underwent both modernisation and improvement programmes – Yasur 2000 in the 1990s and Yasur 2025 at the time of writing. In 1994 the squadron's helicopters infiltrated special forces into Lebanon. The force kidnapped one of the leaders of the Amal terror organisation to receive information regarding the fate of Ron Arad, the IDF/AF navigator who was posted MIA in 1986 over Lebanon.

The worst accident in the history of the Israeli Air Force occurred during the night of 4 February 1997 when two CH-53s from the squadron, fully loaded with infantry, collided in mid-air on their way to Lebanon. The helicopters crashed with the loss of 73 soldiers and crewmembers. During the 2006 Second Lebanon War the squadron was involved in the infiltration of special unit to raid the Hezbollah HQ at Ba'al Beck in Lebanon. The force killed 20 terrorists and collected intelligence information from the HQ. Another operation involved the evacuation of soldiers from a Navy corvette that suffered damage when it was hit by an anti-ship missile. In 2012 the squadron received a new CH-53. This was the result of a two-year special project by AMU 22 that restored an airframe retired in early 2000s, in the process bringing this up to the latest Yasur 2025 standard.

Chapter 14

CH-53 Yasur 2025 serial number 048 from No. 118 'Night Predators' Squadron prepares to deploy a team from Unit 669. (Ofer Zidon)

No. 133 'Twin Tail' Squadron

With the ageing of the previous generation of delta interceptors (Mirage, Nesher and Kfir) and the arrival of improved MiG-21 variants, the MiG-23, Mirage 5 and Mirage F.1 within the Arab air forces surrounding Israel, the IDF/AF required an advanced new interceptor. At the end of 1974 an Israeli delegation visited the US and recommended the purchase of 25 F-15A/Bs. No. 133 'Twin Tail' Squadron was formed on 1 April 1976 at Tel Nof and received its first batch of pre-production F-15As on 10 December 1976. The remainder of the F-15s from the original Israeli purchase began to arrive a year later, with deliveries from December 1977 to 1978.

The F-15's baptism of fire took place during Operation Litani in March 1978, in the course of which 24 operational interception sorties were flown. On 27 June 1979 the squadron scored its first aerial victories. The SyAAF scrambled 12 MiG-21 fighters against four F-15s, which shot down four of the Syrian MiGs, including the first F-15 kill anywhere in the world. Another 'first' may have been achieved on 13 February 1981, with the claimed shooting down of a Syrian MiG-25. The MiG was allegedly

Carrying the individual name *Boomerang*, F-15A serial number 689 from the 'Twin Tail' Squadron manoeuvres after take-off.
(Ofer Zidon)

scrambled to intercept F-4s under way on a high-altitude photo-reconnaissance mission over Syria. The Phantoms dropped down to around 30,000ft (9,144m) with the MiG-25 following them, in turn allowing the F-15 to fire three AAMs and shoot the MiG-25 down. However, the SyAAF had no operational MiG-25s at this time, and the F-15 pilot in question never saw his target.

A subsequent procurement contract brought new F-15s to the squadron from mid-1981. The second batch consisted of nine single-seat F-15Cs and six two-seat F-15Ds that introduced navigators to the squadron for the first time.

On 7 June 1981 six F-15s flew top cover for eight F-16s during Operation Opera, the destruction of the Iraqi nuclear reactor near Baghdad. During the 1982 First Lebanon War the F-15s of the 'Twin Tail' Squadron flew 217 operational sorties with excellent results, shooting down 34 Syrian aircraft with no losses: the squadron's victory tally for the campaign included 17 MiG-23s, 16 MiG-21s and one Aérospatiale Gazelle helicopter.

F-15B serial number 450 lands at Ovda after a training sortie. The aircraft is nicknamed *Cherev Pipiyot* (two-edged sword).
(Ofer Zidon)

Chapter 14

F-15A serial number 663. The aircraft's individual name is *Hamadlik* (the igniter) and the jet is adorned with a single kill marking, after shooting down a Syrian MiG-21 on 27 June 1979. This was the first F-15 kill anywhere in the world. (Ofer Zidon)

During 1983 the squadron began to fly photo-reconnaissance missions across the Middle East. The F-15's superior range and speed allowed it to take some of the burden from the RF-4 fleet.

On 1 October 1985 the unit, together No. 106 'Spearhead' Squadron, commenced Operation Wooden Leg. Eight F-15s from both squadrons flew 2,060km (1,280 miles) to bomb the PLO HQ in Tunis. The sortie is still considered the longest raid executed by the Israeli Air Force.

During the last three decades the squadron has flown numerous operational sorties, including interception and photo-reconnaissance. With no real equal in its air-to-air operations, and with the development of accurate guided munitions, the squadron has also entered the world of air-to-ground operations. Since Operation Accountability in 1993 the squadron's F-15 have flown air defence missions together with ground-attack sorties.

A pair of F-15As performs a synchronised landing on the parallel runways at Ovda. (Ofer Zidon)

No. 210 'Eitan' Squadron

No. 210 'Eitan' Squadron was activated on 20 December 2010, during a ceremony at Tel Nof. The squadron was established in order to operate the latest addition to the IASF's UAV force, the IAI Heron TP, known as the Eitan (firm). This four-ton class, turboprop-powered UAV was designed as a multi-payload, multi-mission, all-weather platform to meet the requirements of the IASF. The Eitan is powered by an 895kW (1,200hp) engine that provides for an operational altitude of 45,000ft (13,716m), above commercial air traffic, and an endurance of 36 hours with a full mission payload.

The acceptance ceremony for the first Eitan UAV took place in February 2011, when IAI delivered the first operational example to the IASF at Tel Nof. Establishing the new squadron at Tel Nof, the largest IASF base in central Israel, symbolises the growing significance of the UAV force within the IASF. Combining the new squadron with the base's older rotary-wing and jet squadrons will help the integration process of the UAV force into the IASF's core operational thinking.

On 29 January 2012 one of the Eitan UAVs crashed as a result of a wing breaking in mid-air. The crash took place shortly after the UAV had taken off from Tel Nof.

Heron TP serial number 210 from No. 210 'Eitan' Squadron at Tel Nof.
(Ofer Zidon)

A Heron TP operator at his station. The upper console shows a moving map display and the lower console presents the UAV's in-flight data.
(Ofer Zidon)

Chapter 14

Heron TP serial number 218 on display at Tel Nof. Note the impressive tail art and the surveillance pod under the forward fuselage. (Ofer Zidon)

Airborne Rescue and Evacuation Unit 669

The IASF's Airborne Rescue and Evacuation Unit 669 usually uses UH-60 or CH-53 assault helicopters for CSAR missions, but also conducts maritime missions using the AS.565MA. A Unit 669 assault team usually consists of soldiers to secure the perimeter around the operation and medical crew to provide the medical treatment required. The unit is also kept busy rescuing civilians in distress throughout the year.

Unit 669 was established in April 1974 as a result of lessons learned from the October 1973 War that highlighted the necessity of a professional combat search and rescue unit. The unit's initial focus was on the extraction and initial medical treatment of pilots downed behind enemy lines, but over the years the unit has also participated in the extraction and evacuation of IDF soldiers and even citizens in distress, for example those caught in winter floods.

The unit's training combines a high level of combat skills with paramedic training. A team is usually based around a combination of soldiers and medical specialists. The exact mix depends upon the location and scale of the event. The unit is trained to work with the CH-53 for large-scale events and the UH-60 for smaller-scale events. The training of the unit's soldiers lasts 18 months and includes:
- Combat medic course
- Parachuting course at the IDF Parachuting School
- Scuba diving course
- Counter-terrorism course at the IDF Counter-Terror Warfare School
- Rappelling course
- Rescue under harsh conditions
- Navigation
- Commanders' course

The unit suffered its most painful loss during the first day of the 1982 Lebanon War, on 6 June, when one of its Bell 212s was shot down by ground fire in Nabatiyeh area, while racing to recover a downed A-4 pilot eventually captured by the PLO. The crew of five was killed. During the first four months of that war, Unit 669 flew 2,518 medical evacuation sorties, recovering battlefield casualties.

In 1997 an assault team from the IDF Naval Commando was operating in Lebanon when it was hit by improvised explosive devices (IEDs), killing 13 soldiers and leaving three wounded. A team from Unit 669 evacuated the Commando's dead and wounded

Paramedics from Unit 669 load a wounded soldier into a CH-53A Yasur 2000. (Ofer Zidon)

Two Unit 669 soldiers in front of a CH-53 Yasur 2000 during an exercise in southern Israel. (Ofer Zidon)

Chapter 14

A Unit 669 soldier is lifted by the winch of UH-60L Yanshuf 2 serial number 830 during a rescue demonstration. A Unit 669 officer observes from the helicopter's front window, while the helicopter's flight engineer commences the lifting process. (Ofer Zidon)

A Unit 669 team rushes to take their posts defending CH-53 Yasur 2000 serial number 036 during a rescue training mission. (Ofer Zidon)

soldiers under intense enemy fire. The team received a citation for its bravery and effectiveness under fire.

In February 2003 the unit rescued 10 Turkish sailors from a ship caught in a storm, despite severe weather conditions including strong winds.

During the 2006 Second Lebanon War the unit evacuated wounded soldiers under enemy fire. In 2008's Operation Cast Lead, one of the unit's medical doctors performed successful emergency surgery in a helicopter during flight, saving the life of a badly wounded infantryman. On 18 January 2010 the unit was scrambled six times to rescue 32 citizens caught in severe winter floods.

Air Maintenance Unit 22

Air Maintenance Unit 22 was established in 1948. At first the unit operated from Sarona in Tel Aviv and was later transferred to Tel Nof, where it operates until today.

AMU 22 is the leading IASF maintenance organisation. The unit is the only one of its kind qualified for D-level maintenance by the aircraft manufacturers – including structural repairs. D-level maintenance is completed once every six to 12 years, depending on the aircraft type. AMU 22 also performs C-level operations, which involve the recov-

AH-64A serial number 956 from No. 190 'Magic Touch' Squadron is stripped down, ready for D-level maintenance at AMU 22 facilities in Tel Nof. D-level work is also used to upgrade the helicopter's systems to the highest standard available at the time.
(Ofer Zidon)

Chapter 14

ery of damaged aircraft after emergency landings or crashes. The unit's specialists participate in investigation committees appointed after every crash of an IASF aircraft.

AMU 22 is based around numerous workshops; each one specialises in a different aspect of aircraft maintenance. Some of the workshops are unique and also serve Israel's civil aviation fleet. While the unit is qualified to perform D-level maintenance on most of IASF assets, exceptions are the relatively small fleets of heavy and light transport aircraft, which receive D-level maintenance from civil companies including IAI Bedek and Elbit Systems. The AMU also provides maintenance services for avionics systems, such as the F-15's central navigation computer.

The AMU takes part in the development of new capabilities for the IASF fleet, including the adoption of new Israeli-made systems and armament. Most of the improvement programmes that have been pursued for Israeli Air Force aircraft over the years were designed and implemented by the AMU. Such projects include the F-4 Kurnass 2000, CH-53 Yasur 2000 and Yasur 2025, F-16C/D Barak 2020 and F-15 Improved Baz and Baz Forever.

The AMU's extensive experience in the maintenance of various aircraft types and its thorough understanding of materials and avionics allow it to initiate special projects in order to rebuild badly damaged aircraft. Two of these projects became airborne in 2012. The first was the reconstruction of a CH-53 that had been retired in the

F-16D serial number 041 was badly damaged when its main undercarriage collapsed during landing. The crew ejected and the aircraft was burnt out. It is seen above after restoration to flying condition by Air Maintenance Unit 22 in a project lasting two and a half years, and (left) making its ferry flight back to No. 109 'Valley' Squadron.
(Ofer Zidon)

205

F-15C serial number 530 undergoes deep (D-level) maintenance with AMU 22 in Tel Nof.
(Ofer Zidon)

early 2000s and was returned to flight condition as replacement for the CH-53 crashed in Romania in 2010. The project lasted for 14 months and brought the old airframe up to the latest Yasur 2025 standard. The second project involved an F-16D that was badly damaged when its main undercarriage collapsed during landing and the aircraft crashed. The restoration project lasted two and half years, and was completed despite the fact that Lockheed Martin had examined the aircraft and stated that it was beyond repair. Eventually, the project cost was equivalent to only 10 per cent of a new airframe and the IASF inventory thus gained one more valuable two-seat F-16D.

MANAT (Flight Test Center)

The IAF Flight Test Center is the unit responsible for flight and weapons testing, airframe modification and avionics integration. MANAT was formed in 1978 and is based at Tel Nof.

MANAT operates a flying unit with test pilots and flight-test engineers, a technical unit in charge of aircraft maintenance, an avionics unit and a UAV unit that operates from Palmachim. The Center operates a fully functional example of every IASF frontline combat aircraft type. It also operates a number of UAVs and will utilise helicopter and transport types from their original squadrons when necessary.

At the time of writing, MANAT operated four F-16 variants: F-16C Block 30 serial number 301, F-16D Block 30 serial number 020, F-16D Block 40 serial number 601 and F-16I serial number 401. Other aircraft operated by the squadron comprised F-15I serial number 201 – the very first F-15I built – and F-15D serial number 714. MANAT has also participated in the evaluation of foreign aircraft types, including the Syrian MiG-23 whose pilot defected to Israel in October 1989 and a pair of MiG-29s loaned from the Polish Air Force in 1995. During their stay in Israel, one of the MiG-29s even received the squadron emblem.

Chapter 14

F-15D serial number 714 is operated by MANAT (note the insignia on the tail). The aircraft's individual name is *Nesher Habarzel* (iron eagle). The aircraft is armed with one of the Ankor series of ballistic missile simulators, used for Arrow anti-ballistic missile trials. The F-15 launches the Ankor, which than simulates the performance and route of a 'Scud' or other ballistic missile en route to attack Israel. (Ofer Zidon)

F-16D serial number 601 operated by MANAT. The Barak is seen during tests with the Rafael Spice guided munition. (Ofer Zidon)

207

An upgraded Yasur 2025, CH-53A serial number 985 from No. 118 'Night Predators' Squadron, makes its maiden flight with the squadron. This helicopter was struck off charge in the early 2000s before being rebuilt during a two-year project by AMU 22 at Tel Nof. The inscription on the engine cowling reads 'Yasur 2025'. (Ofer Zidon)

Chapter 15

AIR DEFENSE COMMAND

On 11 January 2011 the IASF's Anti-Aircraft Command changed its name to Air Defense Command. The new IASF air defence doctrine, which was coupled with the change of name, stipulates that the Command is responsible for the air defence of Israel's home front, complementing the defence of Israel's airspace that is provided by IASF squadrons. The Command is therefore the last line of defence against enemy aircraft threats and the first line of defence against enemy missile threats. ADC is headquartered in Tel Aviv alongside IASF headquarters.

During the last decade missiles threats have developed to pose a strategic threat to the security of Israel. Decades of failure fighting against the IDF drove enemy nations and combatants to equip themselves with thousands of missiles and rockets with reaches ranging from a few kilometres to 1,500km (932 miles) and more. Beside nations such as Syria and Iran, the Gaza-based Hamas and Lebanon-based Hezbollah terror organisations also threaten Israel's northern and southern borders with thousands of rockets of various ranges. In light of this situation Israel has adopted a three-tier defence against the missile and rocket threat.

The higher layer is designed to meet the ballistic missile threat posed by nations such as Syria and Iran, and is based on the IAI Arrow (Chetz) anti-ballistic missile. The development of the Arrow ABM began in the early 1990s and today it is considered to be a mature and operational system with dozens of successful tests behind it. The latest version of the ABM is the Arrow Block 4 with improved range and accuracy.

The second layer is designed to defend Israel against medium-range missiles and rockets. Currently it uses hardware in the form of the Raytheon Patriot PAC-3 (known locally as Yahalom – diamond) with range of 60 to 150km (37-93 miles). The Patriot was used in action for the first time during the 1991 Gulf War, targeting Iraqi 'Scud' missiles launched towards Israel, with moderate success. Improvements made to the system in the last decade are intended to improve its accuracy and detection abilities. Israel considers the PAC-3 a stopgap solution until the newly developed Rafael/Raytheon Magic Wand (Sharvit Ksamim) system is operational. In 2006 Rafael received a contract for the development of this system, with the main requirements including a range of 250km (155 miles), to allow the interception of medium-range missiles and rockets – the primary threat posed against the population of northern and central Israel by Hezbollah.

The lower-layer system, the Rafael Iron Dome (Kipat Barzel), was used in action for the first time in early 2012, with notable success. The Iron Dome is the IASF's latest mobile air defence system and is designed to operate against short- to medium-range rockets and missiles with ranges of 4 to 70km (2.5 to 43.5 miles). Developed over a

Modern Israeli Air Power

period of six years, the Iron Dome draws upon lessons learned from the approximately 4,000 missiles and rockets launched at Israel by Hezbollah during the 2006 Second Lebanon War.

In March 2012 three batteries were deployed around the Gaza Strip, defending the major cities in southern Israel: Beer Sheba, Ashkelon, Ashdod, Ofakim and Netivot. Since then two more batteries have become operational and are also deployed in the area. Terror organisations based in Gaza launched over 200 rockets of which the system identified 68 as potential threats to populated areas, and successfully intercepted 55 of these – a success rate of over 80 per cent.

The second extensive use of the Iron Dome took place during Operation Pillar of Defence in November 2012.

The establish of Air Defense Command in 2011 allows the IASF to focus its efforts to develop an operational answer to missile/rocket threat and emphasises the high importance in which Israel sees a formidable solution to this threat.

A sixth Iron Dome battery, the first to be operated by reservists, was fully equipped by September 2013.

An Arrow ABM missile launcher in the upright position. (Ofer Zidon)

Chapter 15

An Arrow ABM is launched during a test.
(Courtesy of Israel Aerospace Industries)

The Patriot PAC-3 launcher vehicle.
(Ofer Zidon)

A Patriot battery deployment.
(Ofer Zidon)

Chapter 15

The Iron Dome launcher.
(Ofer Zidon)

An Iron Dome interceptor is launched during Operation Pillar of Defence in November 2012. (Nechemia Gershuni)

Appendix I

IASF ORDER OF BATTLE, 2013

Hatzerim Air Base

No. 69 'Hammers' Squadron
F-15I

No. 102 'Flying Tigers' Squadron
A-4N, TA-4H/J

No. 107 'Orange Tail' Squadron
F-16I

No. 123 'Desert Owls' Squadron
UH-60A/L

Air Force Academy
Grob 120, T-6A, Bell 206B, AH-1

Aerobatic Team
T-6A

Modern Israeli Air Power

Hatzor Air Base

No. 101 'First Fighter' Squadron
F-16C

No. 105 'Scorpion' Squadron
F-16D

'Simulator' Squadron
F-15, F-16 simulators

Nevatim Air Base

No. 103 'Flying Elephant' Squadron
C-130J (from 2014)

No. 116 'Defenders of the South' Squadron
F-16A/B

No. 120 'Desert Giants' Squadron
707, Seascan

No. 122 'Nachshon' Squadron
Gulfstream V, 550

No. 131 'Yellow Bird' Squadron
C-130E/H

Appendix

Ovda Air Base

Palmachim Air Base

No. 115 'Flying Dragon' Squadron
F-16A, AH-1

No. 124 'Rolling Sword' Squadron
UH-60L

No. 161 'Southern Cobra' Squadron
Hermes

No. 166 'Hermes' Squadron
Hermes

No. 200 'First UAV' Squadron
Heron

'Yanshuf/Yasur Simulator' Squadron
CH-53, UH-60 simulators

Modern Israeli Air Power

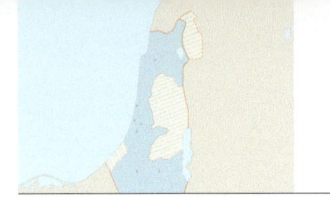

Ramat David Air Base

No. 109 'Valley' Squadron
F-16D

No. 110 'Knights of the North' Squadron
F-16C

No. 117 'First Jet' Squadron
F-16C

No. 193 'Defenders of the West' Squadron
AS.565MA

Ramon Air Base

No. 113 'Hornet' Squadron
AH-64D-I

No. 119 'Bat' Squadron
F-16I

No. 190 'Magic Touch' Squadron
AH-64A

Appendix

No. 201 'The One' Squadron
F-16I

No. 253 'Negev' Squadron
F-16I

Sde Dov Air Base

No. 100 'Flying Camel' Squadron
King Air

No. 135 'Kings of the Air' Squadron
Bonanza, King Air, RC-12

No. 249 'Elad' Squadron
AT-802F

Modern Israeli Air Power

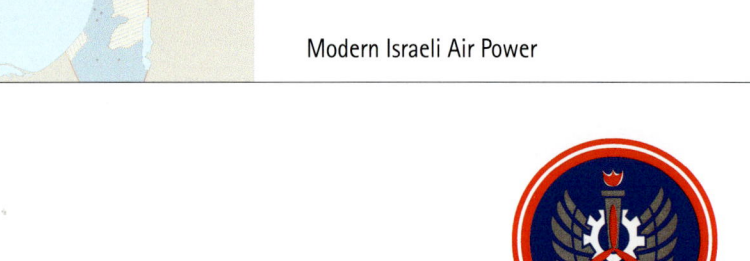

Tel Nof Air Base

No. 106 'Spearhead' Squadron
F-15A/B/C/D

No. 114 'Night Leaders' Squadron
CH-53

No. 118 'Night Predators' Squadron
CH-53

No. 133 'Twin Tail' Squadron
F-15A/B/C/D

No. 210 'The Eitan' Squadron
Heron TP

Airborne Rescue and Evacuation Unit 669
AS.565MA, CH-53, UH-60L

Air Maintenance Unit 22
various types

MANAT (Flight Test Center)
F-15D/I, F-16C/D/I

Air Defense Command   HQ Tel Aviv

HQ Tel Aviv
Arrow

HQ Tel Aviv
Iron Dome

HQ Tel Aviv
Patriot

Appendix II

IASF COLOURS AND MARKINGS

1. Fighters

Boeing F-15I Eagle | *Ra'am (thunder)*

Since its arrival the F-15I has been painted in the usual three-colour camouflage scheme. The need for this camouflage scheme, rather than the air superiority scheme of the rest of the F-15 fleet, is explained by the operational profile of the F-15I – flying low and fast over desert areas. The camouflage colours comprise Matt Pale Stone (F.S.33531), Light Green (F.S.34424) and Red Brown (F.S.30219) over Light Compass Grey (F.S.36375) undersides.

All the F-15I airframes received serial numbers in the 2xx range. Later all received impressive eagle's head tail art. The process of painting the aircraft began in 2004 with a few early sketches before the current artwork was selected and applied to the jets. Another visible addition is the SATCOM antenna dome behind the cockpit. The SATCOM system is a relatively new addition and allows the aircraft to communicate over very long distances.

SATCOM-equipped F-15I tail number 241 from No. 69 'Hammers' Squadron lands at Hatzerim.
(Ofer Zidon)

Modern Israeli Air Power

General Dynamics (Lockheed Martin) F-16A/B Fighting Falcon | *Netz (Sparrowhawk)*

At present the main role of the F-16A/B in the IASF is training. The last remaining front-line Netz squadron, No. 116, is stationed at Nevatim and serves as the OTU and AOTU for new pilots, after they have completed the Air Force Academy and prior to joining one of the front-line squadrons. A second squadron, based at Ovda's Advanced Training Center, provides advanced training for IASF pilots and serves in an adversary role.

The F-16A/B colour scheme has not changed over the years and is based on the standard IASF camouflage of Matt Pale Stone (F.S.33531), Light Green (F.S.34424) and Red Brown (F.S.30219) over Light Compass Grey (F.S.36375) undersides.

F-16A serial number 260 from No. 116 'Defenders of the South' Squadron squadron launches from Nevatim.
(Ofer Zidon)

No. 115 'Flying Dragon' Squadron at Ovda.

No. 116 'Defenders of the South' Squadron at Nevatim.

No. 140 'Golden Eagle' Squadron at Nevatim.

222

Appendix

A close-up of a kill marking (right) and the Osirak raid marking (left). (Ofer Zidon)

The serial numbers of the F-16A/Bs are in the 1xx/2xx range for the single-seat F-16A and in the 0xx/9xx range for the two-seat F-16B. Many IASF squadrons operate one aircraft that bears the squadron number as its serial number, as in the case of Netz serial numbers 140 and 116.

The F-16A/Bs carry prominent tail art, derived from the squadron insignia. No. 116 'Defenders of the South' employs tail art painted in Pale Stone over a Red Brown tail, while No. 115 'Flying Dragon' Squadron uses tail art painted in Black and Red over a Pale Stone tail.

Some F-16s carry Syrian kill markings from the 1982 Lebanon War. Another special marking is painted on the fighters that participated in the Osirak nuclear reactor raid in 1981.

General Dynamics (Lockheed Martin) F-16C/D Fighting Falcon | *Barak (lightning)*

The IASF F-16C/D fleet consists of F-16C Block 30 airframes numbered in the 3xx range, F-16D Block 30 airframes numbered in the 0xx range, F-16C Block 40 airframes numbered in the 5xx range, and F-16D Block 40 airframes numbered in the 6xx range. An exception is Barak serial number 101, which is appropriately assigned to No. 101 'First Fighter' Squadron.

Unique tail art was designed for each of the five F-16C/D squadrons, based on the unit insignia. Two aircraft of the Ramat David squadrons (serial number 364 from No. 110 'Knights of the North' Squadron, as seen on page 40, and serial number 074 from No. 109 'Valley' Squadron) were adorned with kill markings after shooting down Hezbollah UAVs during the 2006 Second Lebanon War.

Modern Israeli Air Power

F-16C serial number 329 from No. 110 'Knights of the North' Squadron armed with a Rafael Spice training munition. (Ofer Zidon)

F-16D serial number 624 from No. 105 'Scorpion' Squadron in an air-to-ground configuration with AIM-9 Sidewinder, Python 5 and JDAM bomb. (Ofer Zidon)

A close-up on the 'Barak 2020' inscription painted on the nose of airframes that underwent the upgrade programme. (Ofer Zidon)

Appendix

No. 101 'First Fighter' Squadron at Hatzor.

No. 105 'Scorpion' Squadron at Hatzor.

No. 109 'Valley' Squadron at Ramat David.

No. 110 'Knights of the North' Squadron at Ramat David.

No. 117 'First Jet' Squadron at Ramat David.

MANAT at Tel Nof.

The colour scheme, like other IASF jet fighters, is based on Matt Pale Stone (F.S.33531), Light Green (F.S.34424) and Red Brown (F.S.30219) over Light Compass Grey (F.S.36375) undersides.

The tail art of the 'Knights of the North', 'Valley' and 'Scorpion' Squadrons is painted in Pale Stone (F.S.33531) over a Red Brown (F.S.30219) tail. The tail art of the 'First Fighter' Squadron is painted in Red Brown (F.S.30219) over a Pale Stone (F.S.33531) tail. The tail art of the 'First Jet' Squadron is painted in Black and Red over Pale Stone tail.

A small inscription was added to the fleet during 2010, to mark participation in the Barak 2020 upgrade programme. The inscription is usually painted under the antenna on the port side of the nose, but some aircraft received it on the starboard side of the intake.

Modern Israeli Air Power

Lockheed Martin F-16I Fighting Falcon | *Sufa (storm)*

The F-16I is the latest Fighting Falcon version in IASF service. The F-16I populates four squadrons: No. 253 'Negev' Squadron was the first to be activated in 2004, followed by No. 119 'Bat' Squadron in 2005, No. 107 'Orange Tail' Squadron in 2006, and No. 201 'The One' Squadron in 2008. Each unit uses its own tail art, representing the squadron's heritage. Serial numbers are predominantly in the 4xx and 8xx ranges. The only exceptions are serial numbers 107 (written off in 2013), 119, 201 and 253, which are assigned to the squadrons that bear these same numberplates.

The colours of the F-16I are identical to those of the F-16A/B and F-16C/D: Matt Pale Stone (F.S.33531), Light Green (F.S.34424) and Red Brown (F.S.30219) over Light Compass Grey (F.S.36375) undersides.

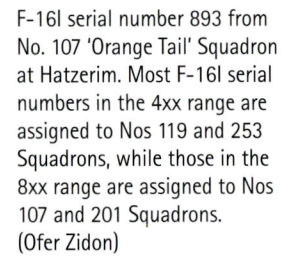

F-16I serial number 893 from No. 107 'Orange Tail' Squadron at Hatzerim. Most F-16I serial numbers in the 4xx range are assigned to Nos 119 and 253 Squadrons, while those in the 8xx range are assigned to Nos 107 and 201 Squadrons. (Ofer Zidon)

No. 107 'Orange Tail' Squadron at Hatzerim.

No. 119 'Bat' Squadron at Ramon.

No. 201 'The One' Squadron at Ramon.

No. 253 'Negev' Squadron at Ramon.

226

McDonnell Douglas (Boeing) F-15A/B/C/D Eagle | *Baz (buzzard)*

The F-15A/B entered Israeli Air Force service in 1976. Very few visible changes have been made to the appearance of the F-15A/B/C/D over its long service. One of changes was the addition of the impressive 'charging eagle' tail art, which was painted on each aircraft that underwent the Improved Baz upgrade programme, in the first decade of the 21st century. Another recent visible addition is the SATCOM antenna dome added to some aircraft just aft of the cockpit.

Each one of the F-15A/B/C/D fighters has a unique individual name applied on the port side below the cockpit. Some of the jets have kill markings and other markings to commemorate their participation in Operation Wooden Leg, the attack on the PLO HQ in Tunis in 1985.

The F-15A/B/C/D fighters retain their original camouflage scheme, based on an overall Light Compass Grey (F.S.36375) colour with patches of Dark Ghost Grey (F.S.36320) on the upper surfaces of the wings, the inner and outer sides of the tails and the radome. Occasionally the radome is painted in a darker shade of grey. The aircraft's unique name and number are painted in dark blue (the same blue as the IASF insignia). All F-15A/B/C/D share the same style of tail art, comprising a swooping eagle motif.

The numbering system of the F-15 is complex, but aircraft were essentially numbered according to delivery batch: Peace Fox I to Peace Fox V. Pre-production F-15As received serial numbers in the 6xx range. Peace Fox I included F-15As (6xx) and F-15B (7xx, later changed to 4xx; 404 and 408). Peace Fox II comprised F-15Cs (8xx) and F-15Ds (7xx, later changed to 9xx and then changed again to 4xx; 450 and 455). Peace Fox III included F-15Cs (5xx) and F-15Ds (2xx; 280). Peace Fox IV comprised F-15Ds (7xx; 706 and 714). Peace Fox V comprised F-15As (3xx; these never entered service) and F-15Bs (1xx; 109 and 113).

F-15D serial number 957 from No. 106 'Spearhead' Squadron. The aircraft name is *Markia Shechakim* (sky blazer) and it is the highest-scoring Israeli F-15, with 4.5 kills. Two MiG-21s and two MiG-23s were shot down in June 1982 and a third MiG-23 was shot down in November 1985, shared with F-15C serial number 840.
(Ofer Zidon)

Modern Israeli Air Power

F-15D serial number 455 from No. 133 'Twin Tail' Squadron. The aircraft's individual name is *Ruach Pratzim* (draught wind). The aircraft carries a special marking (the roundel on the right) to mark its participation in the PLO HQ raid in Tunis in 1985. The kill marking on the left signifies the shooting down of a Syrian MiG-21 on 10 June 1982.
(Ofer Zidon)

A close-up of the tail art on F-15C serial numbers 560 and 828 from No. 106 'Spearhead' Squadron. The markings, common across the 'legacy' Eagle fleet, were applied to mark participation in the Improved Baz upgrade programme.
(Ofer Zidon)

Appendix

2. Transports

Beechcraft Model 36 Bonanza | *Hofit (stint)*

The first batch of Model 36 Bonanzas arrived in Israel in December 2004. The light transport and liaison aircraft replaced the SOCATA TB-20 in the Israeli Air Force's light transport squadron. The colour scheme combines white upper surfaces with grey stripes and grey lower surfaces. Serial numbers are assigned in the 3xx range.

Bonanza serial number 333 from No. 135 'Kings of the Air' Squadron in the static display during an Independence Day open-base event at Tel Nof. (Ofer Zidon)

Beechcraft Model 200 King Air | *Tzufit (sunbird)*

Most of the King Air fleet are painted in an overall grey scheme (F.S.36300). The last two King Airs were purchased from Beechcraft in the mid-2000s and retained their civil colour scheme of overall white with colourful stripes.

The King Airs received serial numbers according to purchase batches: the first batch received serial numbers in the 5xx range, the second batch was numbered in the 6xx range, and the third batch used the 7xx range. The final batch, including the civil-painted aircraft, was allocated serial numbers in the 8xx range.

King Air serial number 848 from No. 135 'Kings of the Air'. These aircraft also serve in the transport stream of the Air Force Academy as trainers for prospective transport pilots and navigators.
(Ofer Zidon)

Modern Israeli Air Power

King Air serial number 859 from No. 135 'Kings of the Air' Squadron. The aircraft combined a civil paint scheme with squadron insignia on the tail. (Ofer Zidon)

Boeing 707 | *Re'em (oryx)*

The IASF tanker fleet is based around the Boeing 707-300 aircraft. Over the years the fleet has changed colour schemes on several occasions, but the last decade saw mainly an overall white scheme, or white upper surfaces with grey underside. Since 2010 the 707 fleet has been undergoing a colour scheme unification process, involving a new overall dark grey scheme, a little darker than that used on USAF tankers. Each aircraft undergoes the painting process while undergoing D-level maintenance in IAI facilities. The 707 fleet received serial numbers in the 1xx (707-320) and 2xx (707-320B/C).

Boeing 707 serial number 295 is one of three already painted in the latest grey scheme. The remaining aircraft will be painted during overhaul. (Ofer Zidon)

Appendix

A close-up of the 'Desert Giants' Squadron insignia on the nose of 707 serial number 264. (Ofer Zidon)

Lockheed C-130E/H and Lockheed Martin C-130J Hercules |
Karnaf (rhinoceros) and Shimshon (Samson)

The C-130 fleet wears the basic three-colour IASF scheme of Matt Pale Stone (F.S.33531), Light Green (F.S.34424) and Red Brown (F.S.30219) over Light Compass Grey (F.S.36375) undersides. Different aircraft wear different markings, according to their mission. Some carry no markings at all, some carry only the IASF roundels, while others carry squadron markings as well. Since both squadrons (No. 103 'Flying Elephant' and No. 131 'Yellow Bird') shared the airframes in the fleet, both squadron

Since both C-130 squadrons pooled their aircraft, squadron insignia were painted on either side of the tail: 'Yellow Bird' insignia on the starboard side and 'Flying Elephant' insignia to port.
(Ofer Zidon)

231

C-130H serial number 420 wears two special markings under the port-side cockpit windows. The markings indicate that it took part in the Entebbe raid in 1976 and in Operation Solomon in 1991, when 15,000 Ethiopian Jews were flown to Israel over one weekend. (Ofer Zidon)

insignia were painted on the tail, with the 'Yellow Bird' insignia on the starboard side and the 'Flying Elephant' insignia to port. As of August 2013, the aircraft had been amalgamated within the 'Yellow Bird' Squadron, while the 'Flying Elephant' Squadron awaited receipt of the C-130J in April 2014.

C-130Es received serial numbers in the 1xx and 2xx ranges, while C-130H serial numbers are in the 3xx, 4xx and 5xx ranges. The new C-130Js are to be painted overall grey with serial numbers in the 6xx range.

3. Attack helicopters

Bell AH-1 Cobra | *Tzefa (viper)*

The AH-1 entered Israeli service in the mid-1970s. Six early-model AH-1G helicopters served on loan with an evaluation unit between 1975–77 before being returned to the US. At its peak, the Cobra fleet contained two squadrons: No. 160 'First Attack Helicopter' Squadron and No. 161 'Southern Cobra' Squadron, both based at Palmachim.

Following the AH-1Gs, five more batches of AH-1s were subsequently delivered to Israel:
- First batch – numbered in the 1xx range and consisted of six AH-1S variants. These aircraft are now retired from service.
- Second batch – AH-1Es numbered in the 2xx range, and arrived from 1978, flying with No. 160 'First Attack Helicopter' Squadron.
- Third batch – AH-1Fs numbered in the 3xx and 4xx ranges, and arrived from 1983. These helicopters formed the basis of the new No. 161 Squadron.
- Fourth batch – AH-1Fs numbered in the 5xx range, arrived during 1990–91.

Appendix

AH-1E tail number 651 from the Air Force Academy's rotary-wing unit, used for advanced attack training. This Tzefa is one of the surplus US Army aircraft that arrived in Israel in 1994. Most of this batch of AH-1s retained the original dark green colour. (Ofer Zidon)

- Fifth batch – numbered in the 6xx range, ex-US Army AH-1Es. Most found their way to the Air Force Academy's rotary-wing advanced attack helicopter squadron.

With the purchase of the advanced AH-64A in the 1990s and AH-64D-I in the 2000s, the AH-1 fleet was downsized to one squadron, No. 160 'First Attack Helicopter' Squadron, before this was disbanded in 2013. AH-1s also serve with No. 115 'Flying Dragon' Squadron, in the adversary role. In 1982 all IDF/AF helicopters were painted in an overall brown colour scheme (F.S.30099 and later F.S.30145). Some helicopters from the fifth batch operated by the Academy retain their original US Army Olive Drab (F.S.34031) colour.

McDonnell Douglas AH-64A and Boeing AH-64D-I Apache |
Peten (python) and Saraf (serpent)

The first AH-64As arrived in Israel in 1990 and comprised a batch of ex-US Army helicopters painted in the US Army dark green colour – US Olive Drab (F.S.34031). The helicopters were divided between No. 113 'Hornet' and No. 190 'Magic Touch' Squadrons. The AH-64A fleet kept its US Army colours until 2005, when a new batch of the advanced AH-64D-I arrived. The AH-64A fleet was amalgamated into the 'Magic Touch' Squadron and the newly built AH-64D-I joined the 'Hornet' Squadron. The AH-64As were re-painted, to match the AH-64D-I colours, in the Israeli desert scheme of Light Stone (F.S.33448) and Brown (F.S.30145) over Light Grey (F.S.36492) undersides.

AH-64As carry serial numbers in the 8xx and 9xx ranges, while the AH-64D-I received serial numbers in the 7xx range.

Modern Israeli Air Power

Seen over the Golan Heights, AH-64A serial number 804 from the No. 190 'Magic Touch' Squadron is armed with Hellfire missiles.
(Ofer Zidon)

AH-64D-I serial number 754 from No. 113 'Hornet' Squadron. The AH-64D-I helicopters all received the desert scheme prior to delivery.
(Ofer Zidon)

4. Assault helicopters

Sikorsky S-65C and CH-53 Sea Stallion | *Yasur (petrel)*

The first batch of CH-53s was delivered to Israel in 1969. Since then, additional helicopters have been assembled from various sources including two ex-Austrian machines and ex-US Marines Corps CH-53As that arrived in two batches in 1974 and in 1991. The fleet has undergone many modifications, including changes to the aircraft numbering system and two improvement programmes: Yasur 2000 in the 1990s and Yasur 2025 in more recent years.

The camouflage scheme has varied over the years, but in the last decade it has been based upon the IASF's standard overall brown helicopter scheme (F.S.30145).

The fleet consists of two squadrons: No. 114 'Night Leaders' and No. 118 'Night Predators' Squadrons based at Tel Nof. The 'Night Predators' Squadron previously adorned its helicopters with unique names below the starboard cockpit door, but with the names were removed as the aircraft underwent the Yasur 2025 programme. A new addition is the border-crossing mission markings below the starboard-side cockpit window. The squadron badge is painted on the starboard side of the tail and on the port-side cockpit door.

CH-53 Yasur 2000 serial number 035 from No. 114 'Night Leaders' Squadron is seen at Tel Nof.
(Ofer Zidon)

Modern Israeli Air Power

CH-53 Yasur 2000 serial number 052 from No. 118 'Night Predators' Squadron lands in the field. The helicopter's individual name is *Ketupa* (fish owl). Note the seven border-crossing mission markings below the window and the 7.62mm (0.3in) machine gun visible through the open window below the winch. (Ofer Zidon)

A close-up on the 'Yasur 2025' inscription. (Ofer Zidon)

Sikorsky S-70A/UH-60 Black Hawk | *Yanshuf (Owl)*

The Israeli UH-60 fleet consists of helicopters that arrived in three batches. The first batch was delivered in 1994 and comprised ex-US Army UH-60As numbered in the 6xx range and painted Olive Drab. The second batch of ex-US Army UH-60Ls arrived in 1999 and was numbered in the 9xx range, once again in Olive Drab. In the early 2000s all the ex-US Army machines underwent the Improved Yanshuf programme in the course of which their serial numbers were changed: 6xx became 7xx, and 9xx became 8xx. The third batch of newly built UH-60L Yanshuf 3s was delivered in 2002. These helicopters were numbered in the 5xx range, and from the outset were painted in the Israeli desert scheme of Light Stone (F.S.33448) and Red Brown (F.S.30219) over Light Grey (F.S.36492) undersides. The older UH-60A/Ls retained their US-style Olive Drab colours until the arrival of the Yanshuf 3, after which some of them received the desert scheme.

Appendix

UH-60A Yanshuf 1 tail number 701 (formerly 601) from No. 123 'Desert Owls' Squadron. The helicopter still flies in its original dark green colours. (Ofer Zidon)

UH-60L Yanshuf 2 serial number 840 (formerly 940) from No. 124 'Rolling Sword' Squadron. This photo shows the locally applied desert camouflage scheme. (Ofer Zidon)

Modern Israeli Air Power

UH-60L Yanshuf 3 serial number 539 from No. 123 'Desert Owls' Squadron, wearing the factory-applied Israeli desert scheme colours.
(Ofer Zidon)

A close-up of the operational sorties markings.
(Ofer Zidon)

Appendix

5. Special missions

Air Tractor AT-802F | *Matar (rain)*

Operated by No. 249 'Elad' Squadron, the fleet of Air Tractor AT-802Fs consists of eight aircraft, originally configured as six with wheeled undercarriage and two equipped with floats. The aircraft wear the standard factory-applied colour scheme of overall yellow, with a prominent blue cheatline extending the length of the fuselage and across the tail. The aircraft are on the Israeli civil register as 4X-AFA, AFS, AFT, AFU, AFV, AFW, AFX and AFY. The two floatplanes were registered as 4X-AFU and 4X-AFX, but these have since been converted to conventional configuration, with wheeled undercarriage. Recently, large, single-digit serial numbers have been added on the aircraft's tails.

Another of the recently acquired AT-802Fs, 4X-AFY/4 is seen at Meggido airfield. The tail numbers were applied some time in 2012.
(Erez M. Sirotkin)

Beechcraft RC-12 | *Kukiya (cuckoo)*

The RC-12 fleet is painted in an overall grey scheme (F.S.36300) and the aircraft are allocated serial numbers in the 9xx range.

RC-12 serial number 990 from No. 135 'Kings of the Air' Squadron. The RC-12s were the first of the King Air family to join the Israeli Air Force. The extensive array of antennas, including the large antenna at the tail, makes it easy to differentiate from the standard King Air.
(Ofer Zidon)

Modern Israeli Air Power

Eurocopter AS.565MA Panther | *Atalef (bat)*

In 1994 the IDF/AF replaced its old HH-65s with the AS.565MA Panther. The five new helicopters were painted in a new scheme of Blue (F.S.25185) over Light Compass Grey (F.S.36375) with distinctive 'bat' tail art. The helicopters are numbered in the 89x and 88x ranges.

AS.565MA serial number 895 from No. 193 'Defenders of the West' Squadron flying over the Mediterranean.
(Ofer Zidon)

Gulfstream V and 550 | *Nachshon Shavit (pioneer – comet) and Nachshon Eitam (pioneer – fish eagle)*

The Nachshon family is based on Gulfstream's business jets, converted to the special missions role in two different configurations. Both types – the Gulfstream V and 550 – have different fairing 'add-ons' to accommodate their mission systems, radars and other electronic equipment. The Gulfstreams initially arrived painted gloss white with blue stripes on both sides of the fuselage and with the serial number and squadron insignia painted on the tail. Over the years some of the aircraft have lost their stripes. Serial numbers are applied in the range of 5xx for the Gulfstream 550 Nachshon Eitam and 6xx for the Gulfstream V Nachshon Shavit.

Gulfstream 550 Nachshon Eitam serial number 569 takes off from Ovda during one of the squadron's deployments.
(Ofer Zidon)

Appendix

Gulfstream V Nachshon Shavit serial number 676 reveals the under-fuselage fairing associated with the ground-scanning radar. (Ofer Zidon)

IAI 1124N Seascan | *Shachaf (seagull)*

The Seascan is based on the IAI Westwind I business jet. The aircraft's paint scheme is overall grey (F.S.36300), similar to that of the RC-12 and King Air. The IASF insignia is applied on the engine cowlings. The aircraft carry no squadron insignia. The nose of the aircraft is painted black. As of summer 2013, just three Seascans remained in operation, including serial numbers 927, 929 and 931.

Seascan serial number 927 from No. 120 'Desert Giants' Squadron taxis in front of one of the squadron's Boeing 707s. (Ofer Zidon)

241

7. UAVs

Elbit Hermes | *Zik (spark)*

The smallest of the UAVs in the fleet, the Hermes 450 is painted in the IASF's standard camouflage colours of Matt Pale Stone (F.S.33531), Light Green (F.S.34424) and Red Brown (F.S.30219) over Light Compass Grey (F.S.36375) undersides. The Hermes are based at Palmachim, where they are operated by two squadrons: No. 161 'Southern Cobra' and No. 166 'Hermes'. Of these, the latter applies prominent tail art on its Hermes, consisting of a swooping black bird of prey, adapted from the squadron insignia. Those Hermes without tail art are operated by No. 161 Squadron, which applies only the squadron insignia.

Hermes 450 serial number 405 at Palmachim. The photo shows the tail art applied on the Hermes UAVs operated by No. 166 'Hermes' Squadron. (Ofer Zidon)

Appendix

IAI Heron | *Shoval (trail)*

The Heron is painted in overall Light Compass Grey (F.S.36375).

Heron serial number 273 from No. 200 'First UAV' Squadron takes off from Palmachim. (Ofer Zidon)

IAI Heron TP | *Eitan (firm)*

The latest component is the Heron TP operated by No. 210 'Eitan' Squadron at Tel Nof. In common with its Heron predecessor, the Heron TP wears an overall Light Compass Grey (F.S.36375) scheme. The 'Eitan' Squadron has applied tail art on the vertical stabilisers of the UAV, comprising a white eagle's against a black background, with red trim.

Heron TP serial number 217 from No. 210 'Eitan' Squadron on display at Tel Nof. (Ofer Zidon)

Modern Israeli Air Power

8. Trainers

Beechcraft T-6A Texan II | *Efroni (lark)*

The T-6 is the latest edition to the Air Force Academy inventory and is the replacement for the veteran Magister. The Texan II has assumed all of the Magister's roles including as mount for the Aerobatic Team.

The T-6 is painted in the Academy training scheme of Gloss White (F.S.17875) and Insignia Red (F.S.11136). The aircraft have serial numbers in the 4xx range.

T-6 serial number 493 is one of the aircraft of the IASF Aerobatic Team.
(Ofer Zidon)

Bell 206B JetRanger | *Sayfan (avocet)*

After the establishment of a second UH-60 squadron in 2002, the Bell 206-equipped No. 125 'Light Helicopter' Squadron at Sde Dov was decommissioned. Some of its helicopters were handed over to the Air Force Academy's rotary-wing basic training squadron, while others found their way to the Israeli police helicopter squadron and some were sold to foreign customers. The helicopters have serial numbers in the 0xx and 1xx ranges.

During 2006, the Bell 206 serving with the Academy's rotary-wing basic training squadron were repainted in a new scheme, based on the Academy's insignia red and gloss white colours.

Bell 206 serial number 125 from the Air Force Academy on display in Hatzerim.
(Ofer Zidon)

Appendix

Grob 120 | *Snunit (swallow)*

The Grob 120 replaced the Super Cub in the initial screening role during the early stages of the IASF flying course. As owner of the fleet, Elbit Systems sells flying hours to the IASF and for this reason the aircraft wear a civil registration with the 4X- prefix. However, some aircraft do carry serial numbers in the 9xx range. The Grob 120 is painted in the Academy colours of gloss white and Day-Glo orange.

Grob 120 serial number 970 4X-DGK.
(Ofer Zidon)

McDonnell Douglas A-4 Skyhawk | *Ahit (eagle)*

The Skyhawk celebrated its 45th year in IASF service in 2012. Once the backbone of the Israeli attack fleet, it is now responsible for advanced training of air cadets with the Air Force Academy. The A-4 fleet has been reduced to one squadron and will soon be replaced by the M-346. The Skyhawk is painted in the standard IASF camouflage

A-4N serial number 102 taxis to the runway at Hatzerim. The aircraft is the squadron's flagship and as such now carries the squadron identity as its serial number. Its original serial number was 332.
(Ofer Zidon)

A close-up of the tiger tail art. (Ofer Zidon)

scheme with the addition of tail art in the form of a large winged tiger and an inscription that indicates that the aircraft underwent under an upgrade programme in 2005.

The current A-4 colours are Matt Pale Stone (F.S.33531), Light Green (F.S.34424) and Red Brown (F.S.30219) on the upper surfaces, and Light Compass Grey (F.S.36375) on the undersides.

The tiger artwork is painted in Brown (F.S.30219) over Pale Stone (F.S.33531). A-4 serial numbers are assigned according to variant: A-4Ns are numbered in the 3xx–4xx range, the TA-4H received serial numbers in the 5xx range and the TA-4Js are in the 7xx range.

Appendix III

CURRENT HEBREW TYPE NAMES

Aircraft type	Hebrew name	In Hebrew	Translation
Fighters			
F-15I	Ra'am	רעם	Thunder
F-16A/B	Netz	נץ	Sparrowhawk
F-16C/D	Barak	ברק	Lightning
F-16I	Sufa	סופה	Storm
F-15A/B/C/D	Baz	בז	Buzzard
Transports			
Bonanza	Hofit	חופית	Stint
King Air	Tzufit	צופית	Sunbird
707	Re'em	ראם	Oryx
C-130E/H	Karnaf	קרנף	Rhinoceros
C-130J	Shimshon	שמשון	Samson
Attack helicopters			
AH-1	Tzefa	צפע	Viper
AH-64A	Peten	פתן	Python
AH-64D-I	Saraf	שרף	Serpent
Assault helicopters			
S-65C/CH-53	Yasur	יסעור	Petrel
S-70A/UH-60	Yanshuf	ינשוף	Owl
Special missions			
AT-802F	Matar	מטר	Rain
RC-12	Kukiya	קוקיה	Cuckoo
AS.565MA	Atalef	עטלף	Bat
Gulfstream V	Nachshon Shavit	נחשון שביט	Pioneer – Comet
Gulfstream 550	Nachshon Eitam	נחשון עיטם	Pioneer – Fish eagle
Seascan	Shachaf	שחף	Seagull
UAVs			
Hermes 450	Zik	זיק	Spark
Heron	Shoval	שובל	Trail
Heron TP	Eitan	איתן	Firm
Trainers			
T-6A	Efroni	עפרוני	Lark
Bell 206B	Sayfan	סייפן	Avocet
Grob 120	Snunit	סנונית	Swallow
A-4	Ahit	עייט	Eagle
Air defence			
Arrow	Chetz	חץ	Arrow
Iron Dome	Kipat Barzel	כיפת ברזל	Iron Dome
Magic Wand	Sharvit Ksamim	שרביט קסמים	Magic Wand
Patriot PAC-3	Yahalom	יהלום	Diamond

Modern Israeli Air Power

(Map by James Lawrence)

BIBLIOGRAPHY

Babich, Col V., 'Egyptian Interceptors in War of Attrition', *Istoriya Aviaciy Magazine* (in Russian), Volume 3/2001

Babich, Col V., 'Egyptian Fighter-Bombers in War of Attrition', *Aviaciya i Vremya Magazine* (in Russian), unknown volume

Centre for Military Studies, *The History of the Syrian Army* (in Arabic), (Damascus, 2001_02)

Central Intelligence Agency, *Probable Soviet Objectives in Rearming Arab States*, Special National Intelligence Estimate No. 11-13-67, 20 July 1967 (released in response to FOIA inquiry by the CIA Historical Review Program)

Cohen, Col E., *Israel's Best Defense* (Shrewsbury: Airlife Publishing Ltd, 1993), ISBN 1-85310-484-1

Cooper, T., Weinert, P., with Hinz, F., and Lepko, M., *African MiGs, Volume 2: Madagascar to Zimbabwe*, (Houston: Harpia Publishing LLC, 2011), ISBN 978-0-9825539-8-5

Cooper, T., 'Geheime Helfer im Yom-Kippour Krieg', *Fliegerrevue Extra Magazine* (Germany), Volume 13

Cull, B., Nicolle, D., and Aloni, S., *Spitfires over Israel* (London: Grub Street, 1994), ISBN 0-948817-74-7

Cull, B., Nicolle, D., and Aloni, S., *Wings over Suez* (London: Grub Street, 1996), ISBN 1-898697-48-5

Davies, S., *Red Eagles: America's Secret MiGs* (Oxford: Osprey Publishing Ltd, 2012), ISBN 978-1846039706

Draper, M. I., *Shadows: Airlift and Airwar in Biafra and Nigeria, 1967–1970*, (Hants: Hikoki Publications Ltd, 1999), ISBN 1-902109-63-5

Roth, A., *Ramat David AFB History*

Roth, A., *Chapters in IAF History: Independence War, Sinai War, Six Days War, War of Attrition, Yom Kippur War*, IAF History Branch (Dapei Shchakim)

Shalom, D., *Like a Bolt out of the Blue – Operation Moked June 1967*,
   (Bavir Aviation & Space Publications),
   ISBN 965-90455-0-6

Shalom, D., *Phantoms over Cairo – War of Attrition 1967–1970*,
   (Bavir Aviation & Space Publications),
   ISBN 965-90455-2-2

Tsiddon-Chatto, Y., *By Day, By Night, Through Haze and Fog*, (Ma'ariv Book Guild)

Weiss, R., and Koren, A., *Aircraft of the Israeli Air Force: McDonnell Douglas A-4 Skyhawk*, (IsraDecal Publications)

Weiss, R., and Koren, A., *Aircraft of the Israeli Air Force: F-15 Baz*,
   (IsraDecal Publications)

Weiss, R., and Koren, A., *Aircraft of the Israeli Air Force: F-16C/D Barak*,
   (IsraDecal Publications)

Weiss, R., *Aircraft in Detail: AH-1 Tzefa*, (IsraDecal Publications),
   ISBN 978-965-7220-16-0

Weiss, R., *Aircraft in Detail: AH-64A/D Peten and Saraf*, (IsraDecal Publications),
   ISBN 978-965-7220-15-3

Weiss, R., *Aircraft in Detail: F-15I Ra'am*, (IsraDecal Publications),
   ISBN 965-7220-04-1

Weiss, R., *Aircraft in Detail: F-16I Sufa*, (IsraDecal Publications),
   ISBN 965-7220-02-5

Weizman, E., *On Eagles' Wings: The Personal Story of the Leading Commander of the Israeli Air Force*, (New York: Macmillan Publishing Co., Inc., 1976),
   ISBN 0-02-625790-4

Official publications

IAF Roots, IAF History Branch

IAF between 1949–1956, IAF History Branch

Israeli Air Force official squadron histories:
   'First Fighter', 'Valley', 'Flying Camel', 'Hammers', 'Flying Elephant', 'Scorpion', 'Knights of the North', 'First Jet', 'Golden Eagle', Central Maintenance Unit 22, 'Defenders of the South', 'Rolling Sword', 'Bat, 'Twin Tail', 'Spearhead', 'Orange Tail', 'The One', 'First Attack Helicopter'.

INDEX

Aircraft
  A-4   66, 72, 89–91, 98, 100, 215, 245–247
  AH-1E/F   17, 52, 53, 67, 78, 84, 98, 132, 215, 217, 232, 233, 247
  AH-64A   16, 32, 53, 54, 172, 218, 233, 247
  AH-64D   16, 53, 166–168, 218, 233, 247
  AS.565MA   60, 162, 163, 201, 218, 220, 240, 247
  AT-802F   59, 186, 187, 219, 239, 247
  Blimp   31
  Beechcraft Model 36 Bonanza   48, 184, 186, 219, 229, 247
  Beechcraft Model 200 King Air   49, 182–186, 219, 229, 247
  Beechcraft RC-12   60, 184–186, 219, 239, 247
  Bell 206B   65, 96, 98, 215, 244, 247
  Boeing 707   17, 37, 50, 123–126, 230
  C-130E/H   33, 37, 51, 114–118, 216, 231, 232, 247
  C-130J   17, 33, 51, 118, 216, 247
  F-15A/B/C/D   33, 45–47, 110, 190–192, 197–199, 205, 206, 216, 220, 227, 228, 247
  F-15I   33, 39, 86–88, 110, 206, 215, 220, 221, 247
  F-16A/B   15, 17, 26, 40, 41, 47, 75, 100, 101, 113, 118–122, 130–132, 158, 162, 165, 216, 217, 222, 223, 247
  F-16C/D   16, 33, 42, 43, 71–73, 105–110, 150–168, 205–207, 216, 218, 220, 223–225, 247
  F-16I   16, 24, 29, 33, 43, 44, 91–93, 111, 165, 169–179, 206, 215, 218, 219, 226, 247
  F-35   17, 47, 122
  Grob 120   64, 66, 96–98, 215, 245, 247
  Gulfstream V & 550   16, 22, 26, 61, 126–128, 186, 216, 240, 241, 247
  Hermes   17, 62, 63, 135, 139–144, 217, 242, 247
  Heron   17, 63, 135, 142–144, 217, 243, 247
  Heron TP   64, 135, 190, 200, 201, 220, 243, 247
  IAI 1124N Seascan   62, 124–126, 163, 216, 241, 247
  M-346   17, 64, 66, 91, 98, 122, 245
  S-65C/CH-53   13, 16, 17, 33, 55, 56, 144, 190, 193–197, 201–203, 205, 206, 217, 220, 235, 236, 247
  S-70A/UH-60   16, 26, 33, 55–58, 67, 94–96, 98, 99, 136–139, 144, 145, 190, 201, 203, 215, 217, 220, 236–238, 247
  T-6A   64, 65, 96–98, 102, 215, 244, 247
  V-22   17, 48, 55

Armament (air-launched)
  AGM-114 Hellfire   27, 28, 83, 84, 167–169, 173, 234
  AIM-7 Sparrow   34, 76, 78–80
  AIM-9 Sidewinder   29, 70, 80, 105, 152, 153, 155, 157, 159, 224
  AIM-120 AMRAAM   37, 72–74, 79–81, 170, 171, 174–178
  ATAP CBU   71
  BGM-71 TOW   20, 67, 84, 99, 223
  BRU-61 rack   82, 83
  Delilah   71, 72, 74, 109, 175

251

ECM/EW pods 75, 76
Ehud pod 78, 79
GBU-10/12/16 26, 43, 83, 152, 153, 170, 177
GBU-15 43, 81, 179
GBU-28 82
GBU-31/32/38 82
GBU-39 82, 83
Griffin 72
Guidance pods 74
Intercept loadout (typical) 176
Litening 74
M117 23
Machtselet (see Spike)
Opher 71
Popeye 43, 47, 73, 74, 191, 192
Python 3 46, 69
Python 4 43, 70, 74, 176
Python 5 39, 43, 70, 73, 74, 80, 81, 105, 155, 159, 170, 171, 174–178, 224
QRA configuration 46
Reconnaissance pods 76–78
Samson 72
Spice 43, 73, 92, 105, 107, 157–159, 171, 174, 207, 224
Spike (Machtselet) 52, 53, 58, 78
Tal 1 & 2 74
Whizzard 71

Armament (surface launched)
Arrow (SAM) 209–211, 220, 247
Iron Dome (SAM) 209, 220, 247
Magic Wand (SAM) 209, 220, 247
Patriot (SAM) 209, 220, 247

Bases
Hatzerim 85, 215, 248
Hatzor 103, 104, 216, 248
Nevatim 33, 100, 113, 114, 216, 248
Ovda 37, 129, 217, 248
Palmachim 135, 217, 248
Ramat David 147–150, 218, 248
Ramon 165, 218, 219, 248
Sde Dov 181, 220, 248
Tel Nov 189, 220, 248

Exercises
Anatolian Eagle 33
International exercises 114, 122
Maple Flag 33, 131
Red Flag 33, 131

Operations
Operation Cast Lead 27, 28, 47, 122, 154, 158, 171, 177, 204
Operation Change of Direction 21
Operation Defensive Shield 19
Operation Pillar of Defence 31, 210, 214
Operation Summer Rain 21
Operation Wooden Leg 16, 46, 190, 192, 199, 227

חיל האוויר והחלל הישראלי

Units
- Advanced Training Center   42, 129–133, 217, 222
- Aerobatic Team   102, 215, 244
- Air Defence Command   209–214, 220
- Air Force Academy   52, 53, 64–67, 96–102
- Air Maintenance Unit 22   204, 205, 220
- Airborne Rescue and Evacuation Unit 669   55, 56, 190, 193, 201–204, 220
- MANAT (Flight Test Center)   206, 207, 220, 225
- No. 69 Hammers Squadron   39, 85–87, 215, 221
- No. 100 Flying Camel Squadron   48, 182–184, 219
- No. 101 First Fighter Squadron   42, 104–107, 147, 216, 223, 225
- No. 102 Flying Tigers Squadron   66, 67, 85, 89–91
- No. 103 Flying Elephant Squadron   48, 51, 114, 115, 117, 118, 122, 147, 189, 216, 231, 232
- No. 105 Scorpion Squadron   42, 104, 108–110, 216, 225
- No. 106 Spearhead Squadron   45, 46, 190–193, 199, 220, 227, 228
- No. 107 Orange Tail Squadron   43, 85, 91–94, 148–150, 215, 226
- No. 109 Valley Squadron   42, 148–154, 205, 218, 223, 225
- No. 110 Knights of the North Squadron   42, 92, 114, 148–150, 154–158, 218, 223, 225
- No. 113 Hornet Squadron   27, 28, 52, 54, 103, 104, 165–169, 172, 218, 233, 234
- No. 114 Night Leaders Squadron   55, 56, 189, 190, 193–195, 220, 235
- No. 115 Flying Dragon Squadron   35, 42, 52, 53, 113, 119, 129–132, 189, 217, 222, 223, 233
- No. 116 Defenders of the South Squadron   41, 42, 75, 91, 101, 113, 114, 118–122, 177, 189, 216, 222, 223
- No. 117 First Jet Squadron   42, 92, 148–150, 158–161, 176, 218, 225
- No. 118 Night Predators Squadron   56, 189, 190, 193, 195–197, 208, 220, 235, 236
- No. 119 Bat Squadron   44, 148, 156, 169–171, 189, 218, 226
- No. 120 Desert Giants Squadron   48, 50, 62, 122–126, 216, 231, 241
- No. 122 Nachshon Squadron   48, 61, 126–128, 216, 240, 241
- No. 123 Desert Owls Squadron   20, 58, 67, 68, 94–96, 99, 100, 215, 237, 238
- No. 124 Rolling Sword Squadron   57, 58, 136–138
- No. 131 Yellow Bird Squadron   48, 50, 114, 117, 118, 125, 216, 231, 232
- No. 133 Twin Tail Squadron   32, 37, 45, 46, 123, 190, 192, 197, 198, 220, 228
- No. 135 Kings of the Air Squadron   48, 49, 60, 181, 184–186, 219, 229, 230, 239
- No. 140 Golden Eagle Squadron   22, 26, 41, 42, 91, 101, 113, 114, 119–122, 222
- No. 160 First Attack Helicopter Squadron   52, 53, 135, 189, 232, 233
- No. 161 Southern Cobra Squadron   135, 139–141, 217, 232, 242
- No. 166 Hermes Squadron   63, 139, 140, 217, 242
- No. 190 Magic Touch Squadron   25, 52, 53, 150, 165, 172, 204, 218, 233, 234
- No. 193 Defenders of the West Squadron   60, 139, 150, 162, 218, 240
- No. 200 First UAV Squadron   63, 135, 139, 142, 144, 217, 243
- No. 201 The One Squadron   86, 103, 104, 165, 174, 176, 189, 219, 226
- No. 210 Eitan Squadron   64, 190, 200, 220, 243
- No. 249 Elad Squadron   59, 186, 219, 239
- No. 253 Negev Squadron   29, 34, 43, 113, 120, 165, 176–179, 219, 226
- Simulator Squadron   104, 110, 216
- UAV School   144
- Unit 669 (see Airborne Rescue and Evacuation Unit 669)
- Yanshuf/Yasur Simulator Squadron   135, 144–146, 217

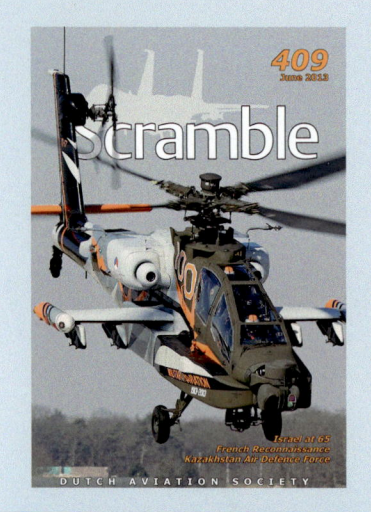

Read all the latest Civil & Military aviation news in

Scramble Magazine

Also available on your tablet or smartphone

As of issue 406 Scramble magazine is in full colour!

Available now:
The Scramble database app
With unlimited database searches & saves

Available on the App Store

Get it on Google play

Now on Sale!

All new edition!

Airfield Guides Western Europe

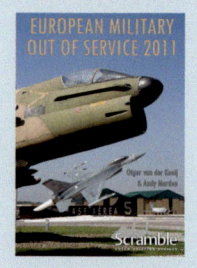

Scramble Airline Fleets Europe 2013

Scramble Military Serials Europe 2013

EMOOS 2011

Our finest publications
DUTCH AVIATION SOCIETY

ACIG

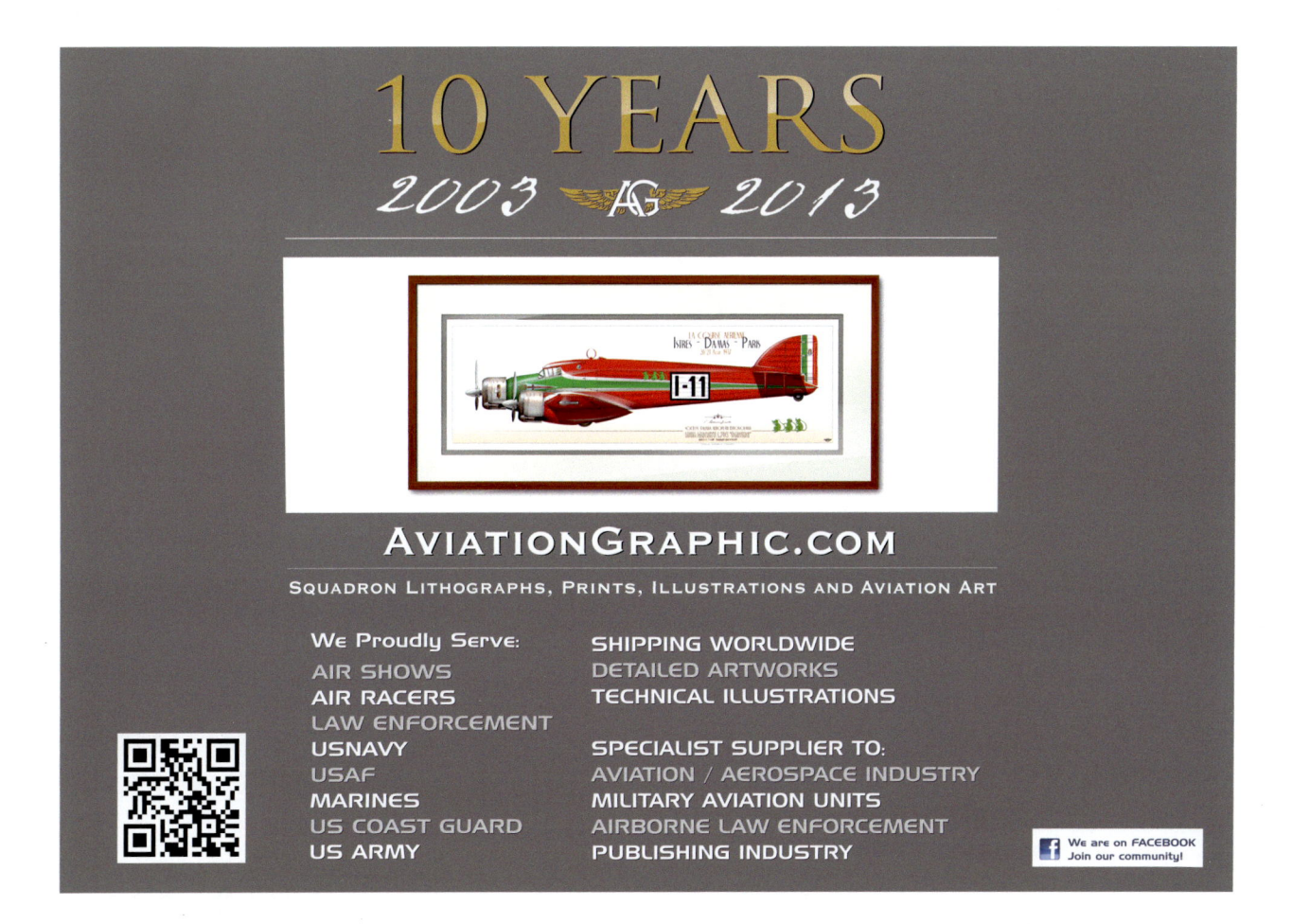

Online since 1999
ACIG is a multi-national project
dedicated to research about
air wars and air forces since 1945

Associated authors, photographers, artists and contributors
have published 24 books, hundreds of articles and artworks.
Multiple research projects are going on and we are
looking forward for your contributions:
join us at ACIG.info forum!

www.acig.info

---

10 YEARS
2003 — AG — 2013

AVIATIONGRAPHIC.COM

Squadron Lithographs, Prints, Illustrations and Aviation Art

We Proudly Serve:
- AIR SHOWS
- AIR RACERS
- LAW ENFORCEMENT
- USNAVY
- USAF
- MARINES
- US COAST GUARD
- US ARMY

SHIPPING WORLDWIDE
DETAILED ARTWORKS
TECHNICAL ILLUSTRATIONS

SPECIALIST SUPPLIER TO:
- AVIATION / AEROSPACE INDUSTRY
- MILITARY AVIATION UNITS
- AIRBORNE LAW ENFORCEMENT
- PUBLISHING INDUSTRY

We are on FACEBOOK
Join our community!

HARPIA PUBLISHING

Glide With Us Into The World of Aviation Literature ...

www.harpia-publishing.com

INTO THE MIDDLE EAST

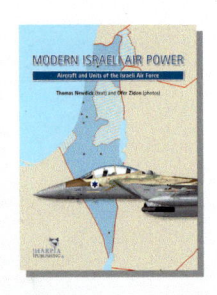

- Modern Israeli Air Power
- IRIAF 2010 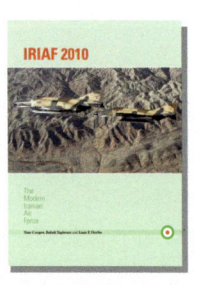
- Iraqi Fighters 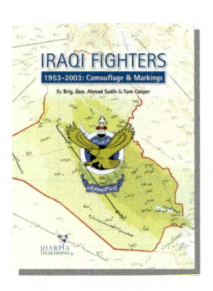

5 YEARS 2008 – 2013

THE ARAB MiGs SERIES

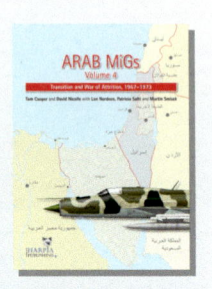

- Arab MiGs | Vol. 4
- Arab MiGs | Vol. 3 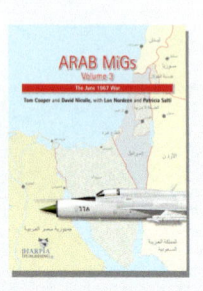
- Arab MiGs | Vol. 2 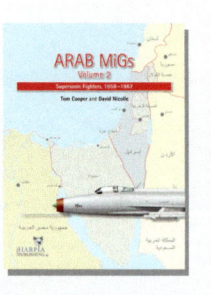
- Arab MiGs | Vol. 1 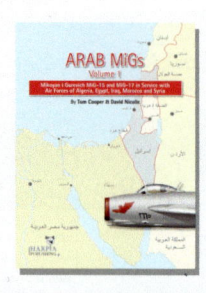

INTO ASIA

- Modern Chinese Warplanes
- Fall of the Flying Dragon 

INTO AFRICA

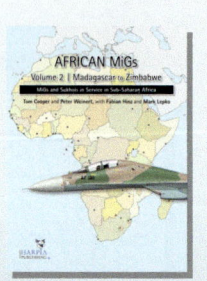

- African MiGs | Vol. 2
- African MiGs | Vol. 1 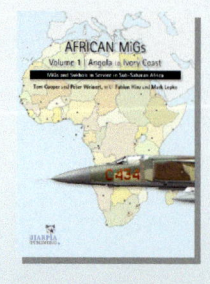

INTO LATIN AMERICA

- Latin American Mirages
- Latin American Fighters 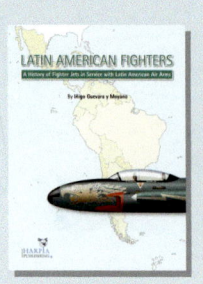

INTO EUROPE

- Silver Wings — K. Tokunaga

VISIT OUR WEBSITE FOR A 16-PAGE FLASH PREVIEW OF ALL OUR TITLES